Biodegradation and Detoxification of Environmental Pollutants

Editor

A. M. Chakrabarty
University of Illinois Medical Center
Chicago, Illinois

CRC Press
Taylor & Francis Group
Boca Raton London New York

CRC Press is an imprint of the
Taylor & Francis Group, an **informa** business

First published 1982 by CRC Press
Taylor & Francis Group
6000 Broken Sound Parkway NW, Suite 300
Boca Raton, FL 33487-2742

Reissued 2018 by CRC Press

Library of Congress Cataloging in Publication Data

Main entry under title:

Biodegradation and detoxification of
environmental pollutants.

Includes index.
1. Microbial metabolism. 2. Xenobiotic
metabolism. 3. Pollution. 4. Chemicals—
Metabolic detoxication. 5. Biodegradation.
I. Chakrabarty, Ananda M., 1938- .
II. Title: Detoxification of environmental
pollutants.
QR88.B56 576'.1133 81-15505
ISBN 0-8493-5524-9 AACR2

A Library of Congress record exists under LC control number: 81015505

Publisher's Note
The publisher has gone to great lengths to ensure the quality of this reprint but points out that some imperfections in the original copies may be apparent.

Disclaimer
The publisher has made every effort to trace copyright holders and welcomes correspondence from those they have been unable to contact.

ISBN 13: 978-1-315-89110-1 (hbk)
ISBN 13: 978-1-351-07020-1 (ebk)

Visit the Taylor & Francis Web site at http://www.taylorandfrancis.com and the
CRC Press Web site at http://www.crcpress.com

PREFACE

The basic objective of this book is to define and review the recent advances in our understanding on the biological mechanisms in the biodegradation and detoxification of various environmental pollutants. Thus, the enzymes involved in the hydroxylation and detoxification of many carcinogenic polycyclic aromatic hydrocarbons by various mammalian tissues have been widely studied, and the genetic basis of such hydroxylation has been delineated. The toxicological problems created by such persistent halogenated hydrocarbons as Kepone® or PBBs have focused attention on the desirability of a detailed study of the biodegradability of such compounds in the environment. Considerable progress in the understanding of the enzymatic breakdown of various chlorinated molecules and pesticides both by aerobic and anaerobic microorganisms, has been made in recent years. Extra chromosomal genetic elements are now known to be involved in the biodegradation of various chlorinated and nonchlorinated hydrocarbons. The involvement of repeated sequences, as well as transposability of plasmid genes, in the evolution of microbial biodegradative functions are also discussed. Finally, various toxic organometallic compounds are released into the environment in the form of fungicides or as by-products of industrial operation. Detailed enzymatic and genetic mechanisms involving reduction or methylation of mercury and its derivatives have been elucidated in recent years. The book encompasses recent studies on the mode of detoxification of such mercurial compounds by various microorganisms.

THE EDITOR

Ananda M. Chakrabarty, Ph.D., is Professor of Microbiology and Immunology at the University of Illinois Medical Center, in Chicago. He was born in Sainthia, India, on April 4, 1938, and received his Ph.D. in Biochemistry from Calcutta University in 1965. He later joined the faculty of the University of Illinois, where he was a research associate and a visiting lecturer in the Biochemistry Division. He joined the staff of the General Electric Company, Corporate Research and Development, in 1971, and has pursued research in genetics of hydrocarbon biodegradation. He has also taught at the State University of New York at Albany, where he was an Adjunct Professor of Biological Sciences. Dr. Chakrabarty received an IR-100 award for constructing multi-plasmid *Pseudomonas* strains for oil spill cleanup and was selected Industrial Research Scientist of the year in 1975. He participated in the Asilomar as well as La Jolla Recombinant DNA conferences for developing guidelines involving recombinant DNA research and acted as a workshop chairman in the National Academy of Sciences forum on recombinant DNA.

Dr. Chakrabarty is a member of the Society of Biological Chemists, American Society for Microbiology, Society of Industrial Microbiology, and American Association for the Advancement of Science.

CONTRIBUTORS

S. Barik, Ph.D.
Post-doctoral Fellow
University of Oklahoma
Norman, Oklahoma

R. W. Eaton, Ph.D.
Department of Medical Biochemistry
University of Geneva
Geneva, Switzerland

Kensuke Furukawa, Ph.D.
Senior Researcher
Fermentation Research Institute
Agency of Industrial Science and Technology
Ministry of International Trade and Industry
Ibaragi, Japan

L. Johnson, Ph.D.
Post-doctoral Fellow
University of Oklahoma
Norman, Oklahoma

Thomas G. Kinscherf, Ph.D.,
Post-doctoral Fellow
Washington University
St. Louis, Missouri

Doug Munnecke, Ph.D.
University of Oklahoma
Department of Botany/Microbiology
Norman, Oklahoma

L. Nicholas Ornston, Ph.D.
Professor of Biology
Yale University
New Haven, Connecticut

D. W. Ribbons, Ph.D.
Professor of Biochemistry and Head of Department
University College of North Wales
Bangor, Gwynedd, United Kingdom

S. Silver, Ph.D.
Professor of Biology, Microbiology, and Immunology
Washington University
St. Louis, Missouri

Henry Talbot, Ph.D.
Post-doctoral Fellow
University of Oklahoma
Norman, Oklahoma

Wu-Kuang Yeh, Ph.D
Senior Research Associate
Department of Biology
Yale University
New Haven, Connecticut

TABLE OF CONTENTS

Chapter 1

MICROBIAL METABOLISM AND ENZYMOLOGY OF SELECTED PESTICIDES

D. M. Munnecke, L. M. Johnson, H. W. Talbot, and S. Barik

TABLE OF CONTENTS

I. INTRODUCTION

The use of synthetic organic chemicals in agricultural practices increased dramatically in the years immediately following World War II. The need to know the fate and effect of these new chemicals in the agricultural environment created tremendous research efforts. Early investigators were primarily concerned with determining which degradative processes were important in establishing the stability of these molecules in the environment. When it was realized that microorganisms played an important role in the degradation of these xenobiotics, more directed biochemical and microbiological projects were initiated to determine the basic principles of microbial metabolism. Thus throughout the 1950s and 1960s, investigators were involved with associating microbial metabolic activities with pesticide transformation. This led some scientists to be amazed at the ability of microbes for degrading these newly synthesized, unique chemicals, claiming that microbes were infallible and could degrade any synthetic molecule, while others were more skeptical and believed that certain molecules could not be metabolized and were therefore considered recalcitrant.[1,2] This debate was tremendously invigorating to the field of pesticide metabolism, since it led investigators toward a better understanding of the basic principles concerning the enzymology and biochemistry of pesticide-related metabolism. This argument also helped to establish the fact that the persistence and fate of these xenobiotics in the environment were strongly affected by microbial activity, and in many situations, microbes were more instrumental in pesticide degradation than were physical or chemical processes.

In the 1970s the mood of the pesticide industry changed from an earlier optimistic feeling to one of recognition of the damages that tremendous large-scale use of these agricultural chemicals could do to our environment as well as to public health. Pesticide usage had increased from 1.1×10^9 lb/year in 1971 to 1.4×10^9 lb in 1977, and to 1.5×10^9 lb by the end of the decade.[3] Problems of chronic exposure,[4] of improper waste disposal,[5] environmental damage,[6] and many more adverse responses to pesticides required that federal laws be promulgated to govern the manufacture, distribution, and use of agricultural chemicals. These laws were aimed at restricting the use of extremely toxic or environmentally dangerous chemicals and reducing the undesirable discharge of pesticides into the environment. In developing these laws, the metabolism studies from the 1950s to 1960s were very helpful in determining which agricultural chemicals, due to their environmental stability and inherent toxicity, truly represented a danger to our chemically dependent society. The new federal laws enacted to control these hazardous chemicals required better technology for control of pesticide waste discharges and for pesticide cleanup. The tremendous diversity in the type of chemicals to be detoxified mandated that many different approaches be considered. Physical methods of incineration, entrapment, and burial were developed,[7,8] as well as chemical methods involving oxidation, reduction, and hydrolysis.[9] However, heavy use of biological treatment systems persisted, and this technology was improved. An important aspect of earlier pesticide metabolism studies helped to develop a new approach to biological pesticide waste treatment technology. Crude enzyme extracts from microorganisms were, in many investigations, able to hydrolyze pesticides to less toxic and less persistent compounds. And the expanding field of biotechnology helped to strengthen the idea that microbial enzymes could be used to control the fate of pesticides in the environment. Using the knowledge gained from research in the 1960s, efforts in the late 1970s led to the demonstration of the potential use of an organophosphate-hydrolyzing enzyme for a variety of detoxification applications, including pesticide detoxification in industrial wastewaters, pesticide containers, and in soils after accidental spills.[10,11]

The intent of this chapter is to review the scientific literature pertaining to the me-

Table 1
TOXICITY OF SELECTED PESTICIDES AND THEIR MAJOR METABOLITES

Pesticide	Major metabolite	Toxicity (mg/kg)[a]	Detoxification factor
Organophosphates			
Paraoxon	—	1.8	
	p-Nitrophenol	350	386
Parathion	—	6	
	p-Nitrophenol	350	122
Dithioates			
Azinphos-methyl	—	13	
	Anthranilic acid	4620	824
Carbamates			
Carbaryl	—	500	
	1-Naphthol	2590	7
Phenylureas			
Linuron	—	1500	
	3,4-Dichloroaniline	100	−23
Monuron	—	1480	
	4-Chloroaniline	300	−8
Acylanilides			
Alachlor	—	1800	
	2,6-Diethylaniline	2690	2
Propanil	—	560	
	3,4-Dichloroaniline	100	−8
Phenoxyacetates			
2,4-D	—	370	
	2,4-Dichlorophenol	580	2
Silvex	—	650	
	2,4,5-Trichloro-phenol	320	2

[a] Toxicity reported: LD_{50} oral, rat, mg/kg, data from Christensen.[12] Dash denotes increase in toxicity.

tabolism of various pesticides by cell-free enzyme extracts in order to show the potential that microbial enzyme systems have in regard to the detoxification and disposal of pesticides in industrial situations. Our basic postulation is that microbial enzymes can be isolated and be effective in converting toxic or persistent pesticides into less toxic or less persistent molecules. For pesticides belonging to the organophosphates, dithioates, carbamates, acylanilides, phenylureas, and phenoxyacetates, hydrolases are instrumental in initial steps of whole cell metabolism. Hydrolases are prime candidates for industrial applications because they are generally stable enzymes and by definition, do not require co-factors. In Table 1, the toxicity of some selected pesticides is listed, as well as the toxicity of their major enzymatic hydrolysis product.[12] For instance, the hydrolysis of parathion to *p*-nitrophenol leads to a 122-fold decrease in overall toxicity and complete removal of parathion. For organophosphates, dithioates, and carbamates, a hydrolysis of the parent pesticide will generally lead to a significant reduction in toxicity. For the other three classes of pesticides shown in Table 1 and discussed in this chapter, the acylanilides, phenylureas, and phenoxyacetates, hydrolysis does not generally lead to a significant reduction in toxicity and sometimes leads to more toxic chemicals. Yet the biospecificity of the pesticide molecule is destroyed, and the metabolites are not as stable in the environment as the pesticide molecule. After the reader has been acquainted with the enzymology of pesticide metabolism, the potential industrial applications of this enzyme technology can be discussed and conclusions be made in regard to the future development of this field of biotechnology.

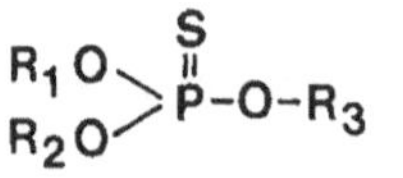

where R_1 and R_2 = alkyl (either ethyl or methyl) moiety

R_3 = substituted aryl or alkyl moieties

FIGURE 1. General chemical structure of organophosphate pesticides.

II. ORGANOPHOSPHATES

A. Metabolism

Pesticides belonging to the organophosphates have recently been extensively used as replacement chemicals for the more persistent organochlorine and mercurial compounds. Excellent reviews have been published discussing the environmental fate of organophosphate pesticides and the role microorganisms play in their degradation.[13-18] The organophosphates are characterized as having a general formula of the type shown in Figure 1. In this class, parathion (O,O-diethyl-O-*p*-nitrophenyl phosphorothioate) is the most extensively studied insecticide due to its intensive agricultural use. Indirect evidence of microbial participation in the degradation of parathion was shown by examining the stability of parathion in autoclaved and nonautoclaved soil and water samples.[19,20] In some cases, heat-labile agents of microbial origin were isolated from the soil, which helped in rapid degradation of parathion. Methyl parathion, like ethyl parathion, also disappeared rapidly in nonsterilized soils but not in sterilized soil.[21] Conclusive evidence for the involvement of microorganisms in parathion degradation was obtained when microbes were isolated from soil, and subsequently in vitro metabolism studies were performed.

Parathion degradation can proceed through basically three different pathways, depending on environmental conditions and the microorganisms involved. Parathion can undergo a nitro group reduction forming amino parathion[18] or an ester hydrolysis forming *p*-nitrophenol and diethylthiophosphoric acid (Figure 2).[18,22] Formation of amino parathion depends upon the organic matter content of the soil, the oxidation-reduction potential, and more specifically, upon the interaction with certain microflora-like yeasts and fungi.[18] The metabolism of parathion to amino parathion creates a more persistent compound.[23,24] Hydrolysis of the phosphoester bond, however, can lead to complete mineralization of the parent molecule. Before hydrolysis, parathion can be oxidized to paraoxon, which can subsequently undergo hydrolysis to *p*-nitrophenol and diethylthiophosphoric acid (Figure 2).

In order to examine microbial metabolism of parathion, various isolation and enrichment techniques were used to obtain microbes capable of metabolizing this compound. Enrichment of microbial populations after repeated applications of parathion to an alluvial soil was demonstrated with the isolation of a bacterial culture (*Pseudomonas* sp. ATCC 29353) capable of using parathion as a sole source of carbon and energy.[25,26] By slight modification of enrichment techniques, bacterial cultures were isolated that could metabolize parathion to diethylthiophosphoric acid and *p*-nitrophenol, or to amino parathion.[27,28] *p*-Nitrophenol was subsequently metabolized by some of these cultures.

Hsieh and Munnecke[29] used a chemostat to isolate microbial cultures capable of detoxifying parathion. A continuous fermentation process was used, and a mixed bacterial culture was adapted to parathion metabolism. None of the bacterial isolates uti-

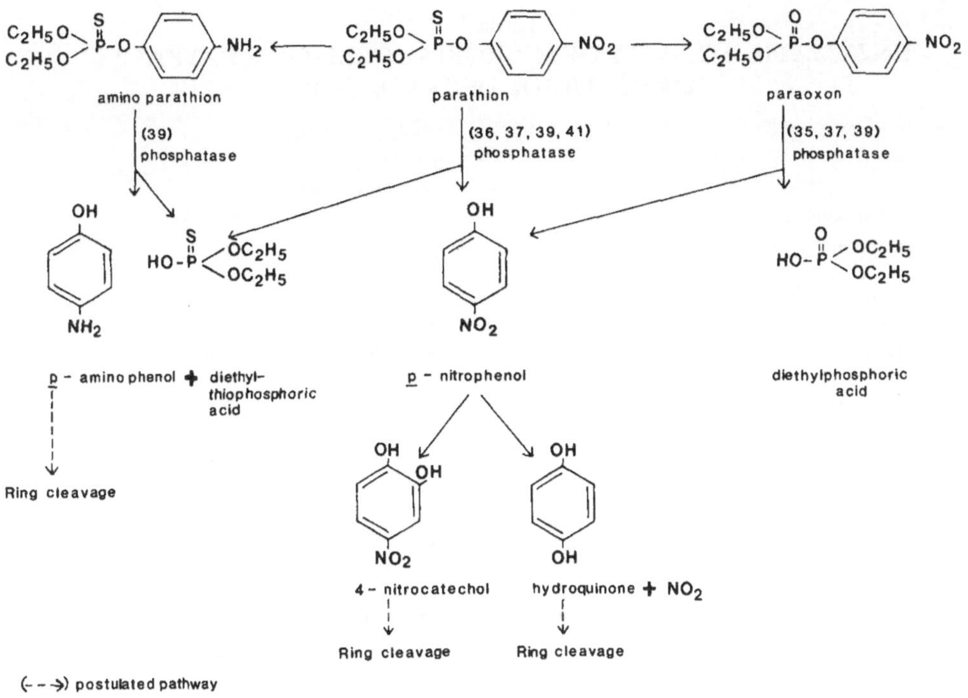

FIGURE 2. Pathways for the metabolism of parathion.

lized parathion as a sole carbon source although eight of the nine cultures utilized *p*-nitrophenol as the sole source of carbon and energy.[30] In a subsequent study, Daughton and Hsieh[31] isolated two *Pseudomonas* species from the mixed culture, *P. stutzeri* which cometabolically hydrolyzed parathion to *p*-nitrophenol, and *P. aeruginosa*, which utilized this nitrophenol as a sole carbon source. Methyl parathion was also metabolized in an analogous manner.

B. Enzymology

In general, the initial point of organophosphate metabolism is the hydrolysis of the arylphosphoester bond. The aromatic metabolites can subsequently be oxidized to produce energy and substrates for cellular growth. Thus this hydrolase (EC 3.1.3), commonly referred to as an esterase, arylesterase, or phosphotriesterase, is the most important enzyme in bacterial metabolism of organophosphates. An analogous enzyme has been extensively studied in both insects and mammals[32] since this reaction is a major mechanism of resistance and detoxification. However, only microbial studies will be reviewed in this chapter. In one of the earliest studies of organophosphate hydrolases, Mounter et al.[33] reported that a group of microorganisms contained enzymes capable of hydrolyzing dialkylfluorophosphates. The DFPase activity (EC 3.7.2.1.) (see Table 2) was found in 10 of 14 tested microbes and was activated by Mn^{2+} and inhibited by Co^{2+}.

The DFPase obtained from *Escherichia coli, Pseudomonas fluorescens, Streptococcus faecalis,* and *Propionibacterium pentosaceum* hydrolyzed fluorophosphate, pyrophosphate, and dimethyl chlorovinyl phosphate compounds at appreciable rates, but failed to hydrolyze dialkylaryl phosphates (trichlorfon, paraoxon), dialkylaryl thiophosphates (parathion, chlorthion) and triethyl phosphates (Table 2).[34]

In another study, Zech and Wigand[35] isolated two phosphohydrolases from *E. coli.* One of the enzymes, DFPase (EC 3.8.2.1) could detoxify DFP (see Appendix) and the

Table 2
ENZYMATIC ACTIVITY OF MICROBIAL CELL-FREE EXTRACTS FOR PESTICIDE DEGRADATION

Pesticide	Enzyme class	Enzyme activity (nmol of substrate transformed/min/mg protein)	Ref.
Organophosphates			
Acephate	Esterase	52	41
Aspon	Esterase	110	41
Cyanophos	Esterase	58	39
Diazinon	Esterase	301, 1200, a	36, 39, 41
Dursban	Esterase	600	39
DCVMP	Lyase	a	34
DEFP	Lyase	a	34
DFP	Lyase	a	34
EPN	Esterase	12	39
Fenitrothion	Esterase	217	39
Fensulfothion	Esterase	238	41
Methyl parathion	Esterase	600, a	37, 39
Monocrotophos	Esterase	133	41
Paraoxon	Esterase	13, 3600, a	35, 37, 39
Parathion	Esterase	259, 3000, 7000, a	36, 37, 39, 41
Propetamphos	Esterase	50	40
Quinalphos	Esterase	1410	40
Triazophos	Esterase	4350	39
TEPP	Lyase	a	34
Dithioates			
Azinphos-methyl	Esterase	87	58
Carbophenothion	Esterase	142	41
Dimethoate	Esterase	62	58
Malathion	Esterase	13, 665, 288, a	41, 56, 57, 58
Malathion	Phosphatase	a	49
Carbamates			
Chlorpropham	Esterase	6	83
Propham	Acylamidase	8	92
Propham	Esterase	5	83
PPG-124	Acylamidase	67	106
Phenylureas			
Carboxin	Acylamidase	252	91
Chlorbromuron	Acylamidase	11, 15	91, 92
Linuron	Acylamidase	18, 20, 130	88, 91, 92
Metabromuron	Acylamidase	16, 18	88, 91
Monalide	Acylamidase	29, 238	91, 92
Monolinuron	Acylamidase	15, 20	91, 92
Monuron	Acylamidase	4	88
Pyracarbolid	Acylamidase	35	91
Acylanilides			
Dicryl	Acylamidase	0.2	96
Karsil	Acylamidase	0.4	96
Propanil	Acylamidase	2, 3, 555	92, 96, 106
Solan	Acylamidase	0.3	96
Phenoxyacetates			
2,4-D	Oxygenase or hydrolase	0.1	114
MCPA	Oxygenase or hydrolase	0.1	127

Note: a = insufficient data to calculate enzyme activity.

other, paraoxonase (EC 3.1.1.2) hydrolyzed paraoxon (Table 2). DFPase could not hydrolyze paraoxon and vice versa. Both enzymes had a very low substrate specificity (K_m = 1.7×10^{-2} and 5×10^{-3} m/ℓ for DFP and paraoxon, respectively), had activity at low pH (8 to 8.5), and were unstable at room temperature.

A cell-free preparation from *Flavobacterium* sp. ATCC 27551 described by Sethunathan and Yoshida hydrolyzed diazinon, parathion, and chlorpyrifos.[36] In all three cases, the aromatic or heterocyclic moiety produced by hydrolysis could not be further metabolized by this bacterium. This phosphoesterase (EC 3.1.3) did not hydrolyze amino parathion or malathion, a dithioate insecticide. In a recent investigation with this *Flavobacterium* sp., Brown[37] reported that the constitutive phosphotriesterase was composed of two protein units, one with a molecular weight greater than 100,000 daltons and the other with a molecular weight of 50,000 daltons. The larger molecule was associated with an orange gum which probably contained flavoprotein. The enzyme was active with those organophosphates having an electron withdrawing aromatic or heterocyclic leaving group as in parathion, paraoxon, and diazinon. The enzyme could not hydrolyze compounds with weakly electrophilic groups like 4-aminophenyl. This phosphoesterase showed activity over a broad range of pH values (8 to 10) and was unaffected by the presence of EDTA, NaF, or NaN_3 or metal ions such as Zn^{2+}, Cu^{2+}, Co^{2+}, Cd^{2+}, Mg^{2+}, or Mn^{2+}. Irreversible enzyme inhibition was seen with non-ionic detergents (0.16%) such as Triton® X-100 and Tween® 80.

In a study by Adhya et al.,[38] this same *Flavobacterium* sp. hydrolyzed both methyl parathion and fenitrothion in addition to diazinon and parathion (Table 2). A *Pseudomonas* species ATCC 29353 isolated from flooded soil[26] was more specific in organophosphate hydrolysis and could hydrolyze parathion and diazinon but not methyl parathion and fenitrothion, two dimethyl aryl thiophosphates. Evidently, the phosphotriesterases from these two organisms are quite different in their specificity even though both hydrolyze the same phosphoester bond.

A crude enzyme extract obtained from a mixed bacterial culture that was grown on parathion showed exceptionally high activity for hydrolysis of at least ten other organophosphates (Table 2).[39,40] The rate of parathion hydrolysis, for example, was 3000 nmol/min/mg protein and was 3000 times faster than chemical hydrolysis by 0.1 *N* NaOH. This phosphoesterase (EC 3.1.3) was unaffected by high substrate concentration, or metabolite accumulation, did not require co-factors, and was active between pH 7.0 to 10.5 with optimal activity at pH 8.5. The temperature optimum of the enzyme was 35°C. This crude enzyme preparation was stable at room temperature in aqueous solution but was denatured by freezing. The presence of various solvents at 15 to 450 ppm accelerated hydrolysis, but at 1000 ppm solvent concentrations the enzymatic activity was reduced. Cleavage of the aryl phosphoester bond was influenced more by alkyl substituents on the phosphorous atom than by functional groups on the aromatic ring. In the case of diazinon, EPN, chlorpyrifos, and triazophos, the effect of aromatic functional groups did not prevent enzymatic hydrolysis, but a $-SCH_3$ group in the para position (fenthion) inhibited hydrolysis. Both dimethyl- and diethyl-substituted dithiophosphates (ethion, azinphos-methyl, chlormephos) were not hydrolyzed by this crude enzyme preparation.

Munnecke[11] covalently bound this crude enzyme extract to porous glass and silica beads and examined the properties of the immobilized enzyme for the detoxification of industrial wastewaters and concentrated pesticide solutions. The immobilized enzyme exhibited approximately 50% of its soluble enzyme activity and displayed minor changes from the characteristics of free enzyme in regard to differences in pH and temperature. An immobilized enzyme reactor could hydrolyze 90% of a 10 ppm parathion solution at flow rates up to 96 ℓ/hr.

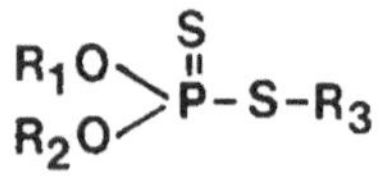

where R_1 and R_2= alkyl (either ethyl or methyl) moiety
R_3=substituted alkyl or aryl moieties

FIGURE 3. General chemical structure of dithioate pesticides.

In another study, Rosenberg and Alexander[41] obtained an inducible phosphoesterase from two *Pseudomonas* species grown in the presence of either diazinon or malathion as sole phosphorous source. A crude enzyme preparation could hydrolyze aspon, monocrotophos, fensulfothion, diazinon, malathion, acephate, parathion, and trithion, but could not hydrolyze dimethoate, trichlorfon, methyl parathion, or dichlorvos (Table 2) (see Appendix). The cleavage of the common phosphorous-oxygen bond of these pesticides was not influenced by the type of alkyl substituents since methyl, ethyl, and propyl esters were metabolized. This enzyme was active against the dithioate insecticides malathion and carbophenothion in contrast to the activity of the parathion hydrolase reported earlier by Munnecke.[39]

All of the above organophosphate hydrolysis studies dealt with the primary hydrolysis of the parent molecule. Clearly this is an important step toward the partial or complete mineralization of these pesticides. For example, *p*-nitrophenol, the major metabolite formed by hydrolysis of parathion, paraoxon, or methyl parathion is readily metabolized by numerous microorganisms, as are other phenol metabolites from other organophosphates.[30,42-45] Cleavage of the aromatic ring after ring hydroxylation involves oxygenases and leads to the formation of intermediates which eventually enter the Krebs cycle. These aromatic pathways have been reviewed elsewhere.[46]

III. DITHIOATES

A. Metabolism

Dithioates constitute a major group of organophosphates. They are characterized by the general chemical formula shown in Figure 3. The pesticides belonging to this group are in great demand due to their broad spectrum and high insecticidal activities. However, not much work has been reported regarding their microbial metabolism except perhaps for one compound, malathion [O,O-dimethyl-S-(1,2-dicarbethoxyethyl)phosphorodithioate].

The hydrolysis of malathion can be both chemical and biological in nature.[47,48] Chemical hydrolysis, as expected, is pH and temperature dependent,[14] but most studies indicate that the major pathway of malathion disappearance in soil, water, sediments, and salt marsh environments is biologically mediated.[47,49-52]

Bacterial cultures isolated from malathion-acclimated estuarine environments degraded malathion either to malathion half-ester or to malathion dicarboxylic acid through half-ester intermediates (Figure 4). One bacterium isolated from a parathion enrichment culture degraded malathion to malathion diacid via half-ester intermediates.[52] The formation of carboxylic acids strongly suggests the activity of carboxyesterase in malathion metabolism.

Malathion disappeared rapidly from salt marsh environments due to bacterial involvement.[49,50] Several bacterial cultures isolated from this ecosystem utilized malathion with or without additional nutrients and formed malathion mono- and dicarboxylic acid, which indicated carboxyesterase enzyme activity. In another experiment, a

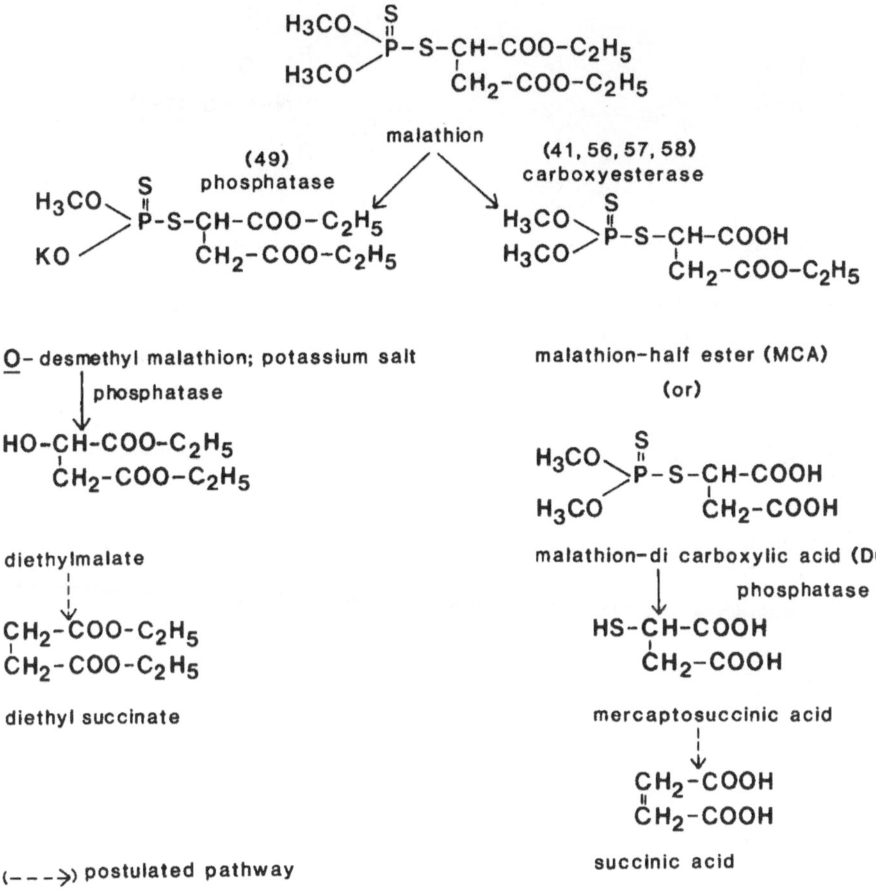

FIGURE 4. Proposed pathway for the metabolism of malathion.

mixed bacterial population utilized malathion as a sole carbon source and formed B-monoacid as the major metabolite.[53] The fungus *Aspergillus oryzae* was isolated from a freshwater pond and also produced B-monoacid and malathion dicarboxylic acid.[54] Both bacterial and fungal cultures had carboxyesterase activity which hydrolyzed the parent molecule to carboxylic acids. These intermediary products were subsequently metabolized to inorganic phosphate and thiophosphates (Figure 4). Demethylation of malathion in bacterial[49,50] and in fungal systems[51,55,56] has been reported. This reaction occurred in some instances after the carboxyesterase reaction, and in one instance demethylation was reported to be a chemical reaction.[48]

B. Enzymology

Two species of the genus *Rhizobium, R. leguminosarum* and *R. trifolii,* were isolated from an Egyptian soil and had powerful carboxyesterase activity (EC 3.1.1.1) toward malathion.[55] Five hydrolytic metabolites, inorganic phosphate and thiophosphate, were identified as metabolites. In a study by Mostafa et al.[51] involving three fungal isolates — *Aspergillus niger, Penicillium notatum* and *Rhizoctonia solani* — the presence of carboxyesterase activity was indicated when malathion was degraded to different carboxylic products which constituted nearly 40% of the total metabolites. Monomethyl phosphate and thiophosphate formed in the culture medium due to the action of two other enzymes, an esterase (EC 3.1.1.2) and a phosphatase (EC 3.1.3). The esterase cleaved the sulfur carbon bond of the mono- and di-acids to dimethyl

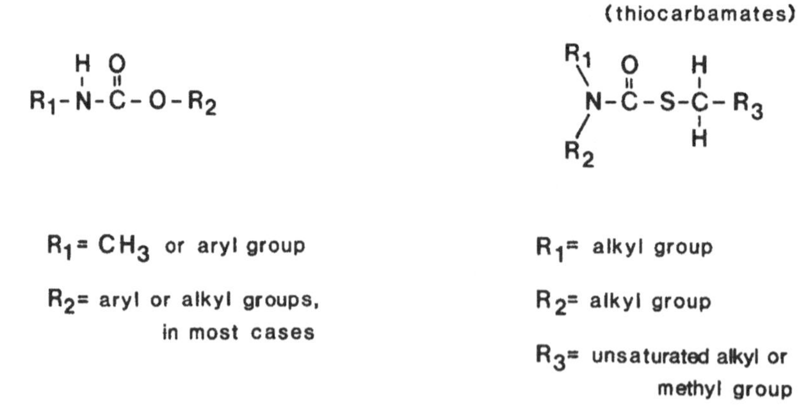

FIGURE 5. General chemical structure of carbamate pesticides, including thiocarbamates.

phosphorothioates and dimethyl phosphorodithioates, which ultimately demethylated through phosphatase activity to these products. This clearly suggests the establishment of these enzymes in cultures metabolizing malathion.[51]

Matsumura and Boush[56] examined a strain of *Trichoderma viride* and a *Pseudomonas* sp. and found that malathion was rapidly metabolized by the action of a soluble carboxyesterase (EC 3.1.1.1), as evidenced by the presence of carboxylic acid derivatives in the culture medium in addition to other demethylated and hydrolytic products.

Cell-free enzymatic hydrolysis of malathion at the phospho-sulfur bond was reported by Rosenberg and Alexander (Table 2).[41] This enzyme, obtained from two *Pseudomonas* species grown on either diazinon or malathion, hydrolyzed carbophenothion and malathion, but failed to metabolize dimethoate. An alkali-extractable, heat-labile organic entity from soil, which transformed malathion to malathion monoacid, was isolated by Getzin and Rosefield[57] from both irradiated and nonirradiated soil. This soil-free, extracellular esterase was quite stable, had a pH optimum of 6.8, and displayed normal Michaelis-Menten kinetics. The enzyme was not denatured with increased acidity (<2.0) and was active over a wide temperature range (20 to 70°C).

A program has been initiated in our laboratory to obtain bacterial cell-free enzymes with specificity for dithioate pesticides. Several bacterial and fungal isolates have been obtained from soil and sewage through enrichment techniques using azinphos-methyl, malathion, or dimethioate as sole carbon sources. Cell-free crude enzyme preparations from the mixed cultures grown on dimethoate could hydrolyze this compound in addition to malathion and azinphos-methyl without any additional salt or co-factors being required.[58]

IV. CARBAMATES

A. Metabolism

The carbamates are a structurally and physiologically heterogeneous group of compounds. Their classification as carbamates is based upon the fact that they are all derivatives of carbamic acid (Figure 5). Some carbamates, such as carbaryl and carbofuran, act as insecticides, whose mode of action is that of inhibiting the enzyme acetylcholinesterase. Other carbamate compounds, such as chlorpropham and diallate, act as herbicides, which control plant growth by interfering with cell division, and additional classes of carbamates act as fungicides and nematocides.[59,60]

The N-methylcarbamates are plant systemic insecticides that are presently in widespread use. One of these compounds is aldicarb [2-methyl-2-(methylthio)propionaldehyde-O-(methylcarbamoyl)oxime]. Coppedge et al.[61] examined the persistence of aldicarb in four different soil types using ^{14}C-labeled compounds and found that the compound was degraded rapidly with the release of radiolabeled carbon dioxide. The half-life of aldicarb under field conditions was observed to be 7 days in a loam soil. Similar results were also obtained from other studies by Andrawes et al.[62] and Richey et al.[63] They found that aldicarb degradation depended on soil type, temperature, and soil moisture. Aldicarb disappeared more rapidly in moist soils. The two major extractable metabolites of aldicarb degradation were reported to be aldicarb sulfoxide [2-methyl-2-(methyl sulfinyl) propionaldehyde-O-(methylcarbamoyl)oxime] and aldicarb sulfone [2-methyl-2-(methylsulfonyl)propionaldehyde-O-(methylcarbamoyl) oxime]. Jones[64] investigated the metabolism of aldicarb by five different genera of soil fungi (*Gliocladium* sp., *Pencillium* sp., *Cunninghamella* sp., *Rhizoctonia* sp., and *Trichoderma* sp.) and discovered that all metabolized the pesticide, with *G. catenulatum* having the most activity. This isolate converted 35% of the added aldicarb (.05 mg) to water-soluble compounds within 21 days. *G. catenulatum* and *C. elegans* also metabolized aldicarb sulfoxide. Aldicarb was not used as a source of carbon and energy, so presumably degradation by these organisms was via a cometabolic process.

Carbaryl (1-naphthyl-N-methylcarbamate), another methylcarbamate insecticide, is also rapidly degraded in soil to polar metabolites. Rodriguez and Dorough[65] used ^{14}C-labeled carbaryl to determine that 72% of the radioactivity in 10 ppm of the pesticide was lost from certain soil samples after 4 days of incubation. A strain of *P. implicatum* was isolated that could hydroxylate carbaryl to 1-naphthyl-N-hydroxy methylcarbamate. In 14 days, 34% of the carbaryl added to the culture medium (1 ppm) was metabolized. Some loss of carbaryl was noted when the pesticide was incubated with strains of *Pseudomonas, Nocardia, Arthrobacter, Xanthamonas,* and *Bacillus.* After 14 days, 90% of the initial carbaryl was degraded in comparison to 40% in an uninoculated control. No strains of bacteria or fungi were obtained that could use carbaryl as a carbon source. Liu and Bollag[66] investigated carbaryl metabolism in the fungus *G. roseum,* which was isolated from soil. Three metabolites were formed by this fungus as a result of hydroxylation reactions; 1-naphthyl-N-hydroxymethylcarbamate, 4-hydroxy-1-naphthylmethylcarbamate, and 5-hydroxy-1-naphthylmethylcarbamate. The first metabolite was formed by the hydroxylation of the side chain and the other two metabolites were formed by the hydroxylation of the aromatic ring. No hydrolysis reactions were detected with carbaryl, but the three metabolites were all less active as acetylcholinesterase inhibitors. Bollag and Liu[67] also isolated fungi from soil that could hydroxylate carbaryl. These included strains of *Aspergillus flavus, A. terreus, Mucor racemosus, Penicillium* sp., and *Rhizopus* sp. The same hydroxylated metabolites as described above were detected in these experiments.

Sikka et al.[68] isolated several genera of marine microorganisms capable of converting carbaryl to water-soluble products within 7 days. Two fungal genera, *Culcitalna* and *Halosphaeria,* were most effective in catabolizing carbaryl, but no hydroxylation of the compound was detected. A chemical and biological hydrolysis product of carbaryl, 1-naphthol, was more susceptible to degradation than was carbaryl. Similar results were obtained by Bollag and Liu[69] with a strain of *Fusarium solani,* which possessed a constitutive enzyme system for the rapid degradation of the naphthyl moiety. A cell-free extract from this organism converted 74% of the radioactivity in ^{14}C-labeled 1-naphthol to radiolabeled carbon dioxide within 30 min. Boush and Matsumura[70] isolated a strain of *Pseudomonas melophthora,* an obligate symbiont of the apple maggot, that had degradative activity against carbaryl. Carbaryl was not utilized as a carbon source, nor was there any evidence of an oxidative attack on the molecule, but water-soluble metabolites were produced. Later, Matsumura and Boush[71] isolated a strain of *Trichoderma viride* that could only weakly degrade carbaryl, but did show evidence of hydroxylation.

Carbofuran (2,3-dihydro-2,2-dimethyl-7-benzofuranyl-N-methylcarbamate) is another important N-methylcarbamate insecticide, Williams et al.[72] studied the microbial degradation of carbofuran by measuring radiolabeled carbon dioxide release from ^{14}C-carbonyl-labeled carbofuran. They observed that the degradation rate was 40 times greater in nonsterile soil as compared to sterile soil. Getzin[73] found that chemical hydrolysis in alkaline soils was the most important mechanism of carbofuran breakdown, but that in acid and neutral soils, both chemical and microbial processes were involved. Yu et al.[74] studied the degradation of carbofuran in a model ecosystem and were able to detect the metabolites carbofuranphenol, 3-ketocarbofuran, 3-hydroxycarbofuranphenol, N-hydroxymethylcarbofuran, and 3-hydroxycarbofuran. Their experiments indicated that N-methylcarbamic acid was formed and was further degraded to carbon dioxide and other compounds. Kandasamy et al.[75] studied a *Helminthosporium* sp. which could degrade greater than 70% of a 20 ppm amount of carbofuran within 14 days. Strains of *A. niger* and *T. viride* had less activity for carbofuran degradation. All three of the isolates formed hydroxycarbofuran as an intermediate.

Benezet and Matsumura[76] screened bacteria and fungi from soil and water to examine the degradation of the methylcarbamate, mexacarbate (4-dimethylamino-3,5-xylylmethylcarbamate). An isolate of *T. viride* was obtained that could decarbamylate the molecule. The major metabolite detected was 4-dimethylamino-3,5-xylenol.

The thiocarbamates are an important subclass of carbamate herbicides which are used on field and vegetable crops. The first herbicide in this class was S-ethyl-dipropyl thiocarbamate (EPTC). MacRae and Alexander[77] measured the microbial degradation of EPTC and observed the release of radiolabeled carbon dioxide from the ethyl moiety of the molecule. Two of the more recently developed thiocarbamate herbicides are diallate [S-(2,3-dichloroallyl)diisopropyl thiocarbamate] and triallate [S-(2,3,3-trichloroallyl)diisopropyl thiocarbamate]. Banting[78] observed that diallate had a half-life in soil of 12 days, whereas triallate was more persistent and had a half-life of 20 days. Anderson[79] used ^{14}C-labeled diallate to examine its degradation and recovered radiolabeled carbon dioxide as the primary metabolite due to the high level of microbial activity in the soil sample. Anderson and Domsch[80] studied diallate degradation in five different soils and reported that 50% of the starting level of 2.5 ppm of diallate was lost in 4 weeks in those soils with an active microbial population. Isolates of *Phoma eupyrena, Penicillium janthinellum* and *T. harzianium* could break down at least 20% of the applied herbicide (2.5 ppm) after 10 days. Of these three microorganisms, only *T. harzianium* was able to use diallate as a supplemental carbon source. If *T. harzianium* was capable of completely degrading diallate, the first step in the pathway may have been hydrolytic cleavage of the ester bond, as Kaufman has postulated.[81]

B. Enzymology

More is known concerning the microbial enzymology of the phenylcarbamate herbicides than any other group of carbamate pesticides. The best characterized carbamate hydrolyzing enzyme was obtained by Kearney and Kaufman[82] when they isolated a strain of *Pseudomonas striata* Chester from soil which could metabolize the phenylcarbamate herbicide, chlorpropham, isopropyl-N-(3-chlorophenyl)carbamate (CPC) as a sole source of carbon and energy. Enzymatic cleavage of the ester and/or amide linkage of chlorpropham was quantitated by the appearance of 3-chloroaniline (Figure 6). The esterase activity (EC 3.1.1.1.) contained in a cell-free extract from the pseudomonad also cleaved propham, isopropyl-N-phenylcarbamate (IPC), as well as nine

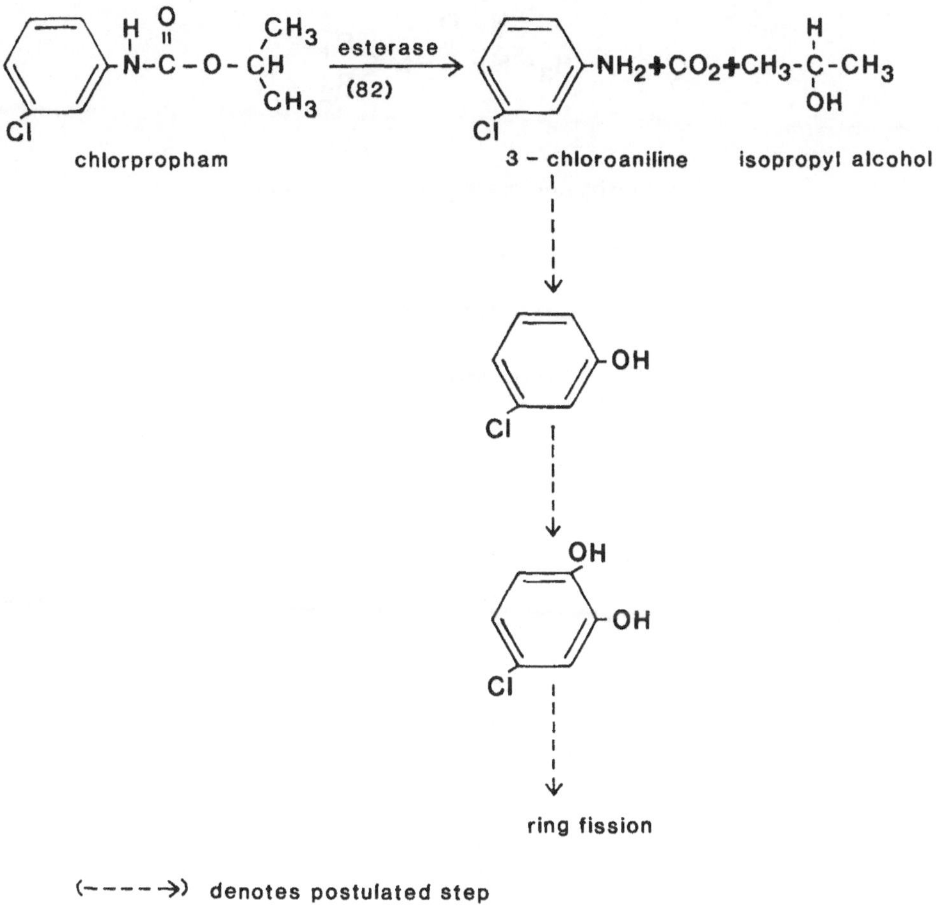

FIGURE 6. Proposed pathway for the metabolism of chlorpropham (CIPC).

other carbamates, including alpha-carbo(2,4-dichlorophenoxyethoxy)ethyl N-(3-chlorophenyl)carbamate, isopropyl-N-(3,4-dichlorophenyl) carbamate, sec-butyl-N-(3,4-dichlorophenyl)carbamate, and alpha-carboisopropoxyethyl-N-(3-chlorophenyl)carbamate. Either 3-chloroaniline, aniline, or 3,4-dichloroaniline was detected from all of the carbamates after esterase activity. Kearney later reported on the purification and characterization of this enzyme.[83] With chlorpropham as the carbon source, the pseudomonad isolate was mass-cultured and the cells then sonicated. Ammonium sulfate precipitation, gel filtration, and ion exchange were used to purify the enzyme. After purification, the pH optimum was 8.5, but the enzyme had a broad range of nearly maximum activity between pH 6.0 and 9.5. With chlorpropham and propham as substrates, the esterase had specific activities of 6 and 5 nmol/min/mg protein, respectively (Table 2). The metal ions Mg^{2+}, Mn^{2+}, Ca^{2+}, and Co^{2+} were not necessary for enzyme activity. The esterase was inhibited 42% with diisopropylfluorophosphate at a 10 to 6 *M* concentration, which implied that the hydroxyl group of serine possibly was part of the active site of enzyme-substrate interaction. The broad substrate specificity of the enzyme was demonstrated by its esterase activity against the carbamates, plus it also showed amidase activity. As an amidase (EC 3.5.1.4), the enzyme hydrolyzed 3,4-dichloroacetanilide and 3,4-dichloropropionanilide. It was proposed that large, bulky substituents in the area of the carbonyl carbon (such as larger alcohol groups) could cause steric hindrance and inhibit the hydrolysis of compounds with this property.

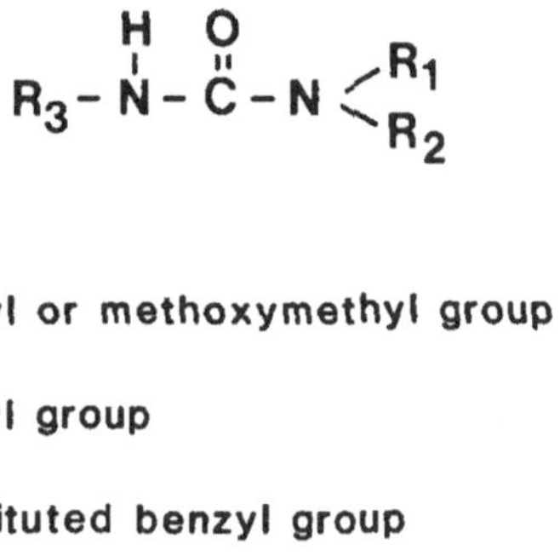

FIGURE 7. General chemical structure of phenylurea herbicides.

V. PHENYLUREAS

A. Metabolism

The phenylureas were first developed after World War II by Du pont®, and approximately 25 urea herbicides are now commercially available. The basic structure of the phenylurea herbicides is shown in Figure 7. The microbial metabolism of urea herbicides has been investigated for many years, and the reader is referred to Geissbuhler et al.[84] for a general discussion of this subject. Only the most salient points of microbiological phenylurea degradation will be presented here.

In one of the earliest studies designed to investigate phenylurea metabolism, Hill et al.[85] reported the decomposition of monuron [3-(*p*-chlorophenyl)-1,1-dimethylurea] and described a *Pseudomonas* sp. that oxidized the herbicide when yeast extract was present in growth media. In a later study, Ross and Tweedy[86] described the metabolism of four phenylurea herbicides by microbial populations from two soils. Two dimethylurea herbicides (fluometuron and chloroxuron) and two methoxymethylphenylurea herbicides (metabromuron and chlorbromuron) were investigated, and it was observed that the dimethyl herbicides were successively N-demethylated by organisms from both soil types, whereas N-demethylation of the methoxymethyl compounds was detected in only one soil type. Only slight N-demethoxylation was observed in all cultures studied by Ross and Tweedy.[86] Instead, the metabolism of metabromuron and chlorbromuron proceeded in some cases via direct hydrolysis to their respective anilines with subsequent transformation to form acetanilide (3-bromoacetanilide).

Weinberger and Bollag[87] described the ability of the fungus *Rhizoctonia solani* to degrade the phenylurea herbicide chlorbromuron [3-(3-chloro-4-bromophenyl)-1-methoxy-1-methylurea]. The demethylated product [3-(3-chloro-4-bromophenyl)-1-methoxyurea] was isolated and identified. In addition, these investigators indicated that the fungus converted the demethylated herbicide to its demethoxylated product, that being the corresponding urea, and this compound was converted to 3-chloro-4-bromoaniline which completely and rapidly disappeared from the culture medium. They also noted that *R. solani* was capable of attacking a large number of other phenylurea herbicides, including chloroxuron, diuron, fenuron, fluometuron, linuron, metabromuron, neburon, and siduron (see Appendix). These herbicides were transformed despite the variations in ring structure or functional groups. The ability of this organism to degrade a variety of phenylurea herbicides suggests the organism may have one enzyme with a relatively broad substrate specificity, but since this study involved only whole cells, it is impossible to predict the exact nature of the enzyme system(s) involved.

B. Enzymology

The degradation of urea herbicides by cell-free extracts of *Bacillus sphaericus* was described by Wallnofer and Bader (Table 2).[88] This organism was reported to inactivate N′-methoxyphenylurea herbicides by liberating carbon dioxide from the ureido portion of the molecule with the subsequent formation of halogen-substituted aniline moieties.[89] Cells of *B. sphaericus* ATCC 12123, when grown in a minimal salts medium supplemented with glucose, yeast extract, and asparagine, were induced to degrade linuron [(N-3,4-dichlorophenyl)-N′-methoxy-N′-methylurea] by the presence of 50 μmol of this herbicide. Induction was not possible in minimal medium containing peptone or meat extract, however. Once cells were induced, cell-free extracts with amidase activity (EC 3.5.1.4) were capable of liberating carbon dioxide from the ureido portion of the molecule, thereby leaving the corresponding aniline moiety. In addition, the crude extract was also capable of decomposing the N′-methoxymethylureas monolinuron [N-(4-chlorophenyl)-N′-methoxy-N′-methylurea] and metabromuron [N-(4-bromophenyl)-N′-methoxy-N′-methylurea], exhibiting specific activities of 14.6 and 16.2 nanomoles of herbicide degraded per minute per milligram protein, respectively. Monuron, a dimethylurea herbicide, was also degraded by the linuron-induced enzyme, which, in this case, exhibited a specific activity of 4.1 nmol/min/mg protein. However, fluometuron, [N-(3-trifluoromethylphenyl)-N′-N′-dimethylurea], also a dimethylurea herbicide, was not attacked by this enzyme. The enzyme expressed a specific activity of 17.8 nmol/min/mg protein for the inducer, linuron (Table 2).

The amidase (EC 3.5.1) responsible for hydrolyzing these herbicides was partially purified by ammonium sulfate precipitation and DEAE column chromatography. Wallnofer and Bader[88] reported a 6.6-fold purification of an enzyme having a pH optimum between 7.0 and 8.5. No enzyme activity was observed below pH 5.5, and the highest enzymatic activity was observed in the late logarithmic-growth phase, prior to sporulation. The authors recognized the relatively low activity of the enzyme system but hypothesized that the enzyme apparently hydrolyzed N′-methoxyphenylureas at the amide linkage, which could yield phenylcarbamic acid and an unidentifiable metabolite. Phenylcarbamic acid spontaneously disintegrated to the corresponding aniline with the formation of carbon dioxide.[83] Subsequently, Engelhardt et al.[90] identified N,O-dimethyl hydroxylamine as a degradation product of linuron. The herbicide was degraded by extracts of *B. sphaericus* ATCC 12123, and the production of N,O-dimethyl hydroxylamine, as well as the products CO_2 and 3,4-dichloroaniline, allowed a mechanism of linuron degradation to be proposed (Figure 8).

In later enzyme-related work with *B. sphaericus*, Engelhardt et al.[91] examined the substrate specificity of the linuron-degrading enzyme and also studied the reaction sequence of linuron degradation. The investigators concluded that linuron, or a compound derived from it, induced an enzyme system responsible for degrading the herbicide. The specificity of the enzyme was quite broad and the acylanilide herbicides monalide, carboxin and pyracarbolid, and several other acylanilide compounds were degraded by the linuron-induced amidase. In fact, the acylanilides were degraded ten times faster than were the methoxy-substituted phenylureas. Because the enzyme exhibited relatively low substrate specificity for both acylanilides and the phenylureas and because the enzyme exhibited a stronger affinity for the acylanilides than for the phenylurea herbicides, Engelhardt and co-workers[91] believed that the site of enzymatic hydrolysis was the peptide bond between the amino group of the aniline moiety and the carboxy group. This led the investigators to classify the enzyme as an acylamidase (EC 3.5.1).

Most recently, Engelhardt et al.[92] described the further purification and properties of the aryl acylamidase (EC 3.5.1.13) of *B. sphaericus*. This enzyme was induced by various phenylamides including acylanilide, phenylcarbamate, and methoxy-substi-

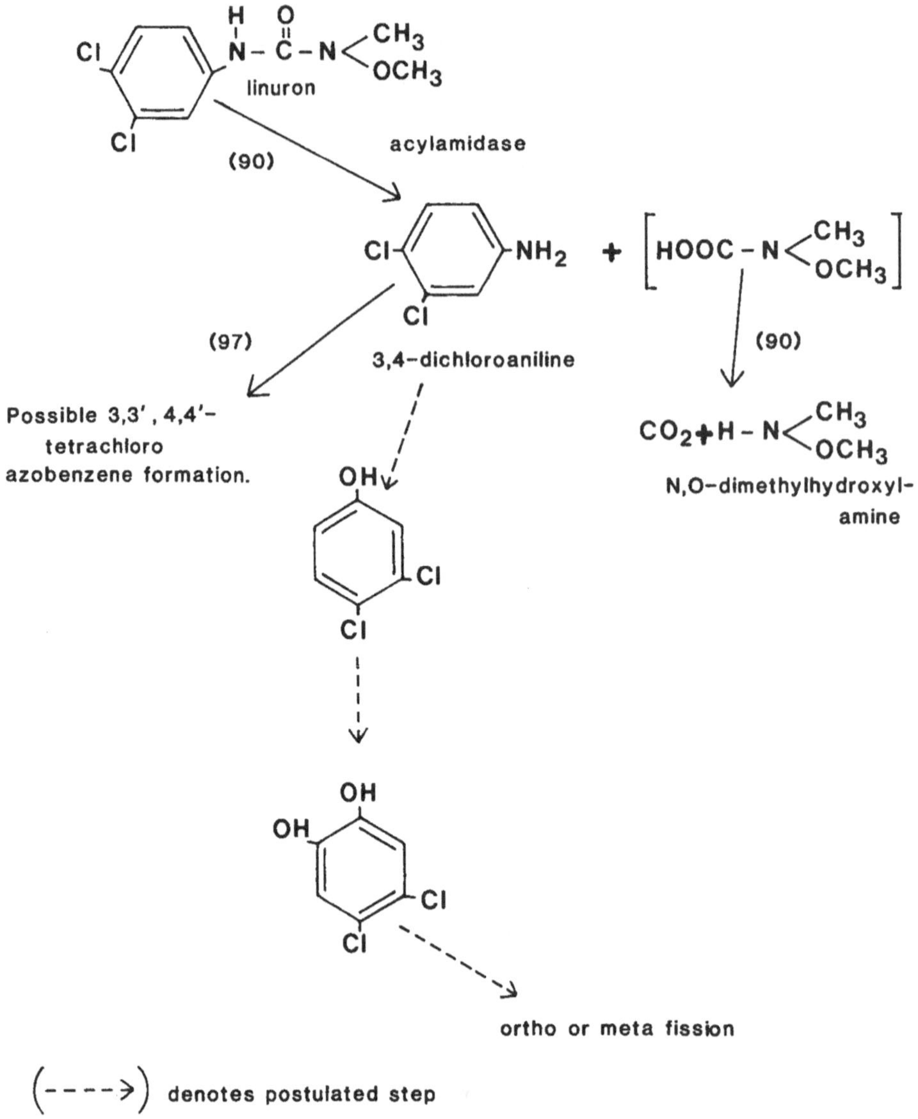

FIGURE 8. Proposed pathway for the metabolism of linuron.

tuted phenylurea pesticides. The results of this study were in contrast to previous studies by Geissbuhler et al.[93] and Tweedy et al.[94] which indicated that a stepwise demethylation or demethoxylation, deamination, and decarboxylation of phenylurea herbicides occurred. The enzyme investigated by Engelhardt et al.[92] resembled the enzyme from *Pseudomonas striata* Chester, studied by Kearney,[83] the acylanilide hydrolyzing enzymes of *Fusarium solani* reported by Lanzilotta and Pramer,[95] and an enzyme from a *Penicillium* sp. described by Sharabi and Bordeleau.[96] These enzymes will be discussed in the phenylurea enzymology section.

The enzyme described by Engelhardt et al.[92] from *B. sphaericus* was produced by cells grown in media where linuron did not serve as the sole source of carbon and energy. Instead, the herbicides were present in growth media at a concentration of 0.05 μmol/mℓ, with glucose, asparagine, and yeast extract serving as the primary carbon substrates, and cells were allowed to grow for 48 hr. When supplied at levels of 0.05 μmol/mℓ, the phenylurea herbicide, maloran [3-(3-chloro-4-bromophenyl)-1-methoxy-

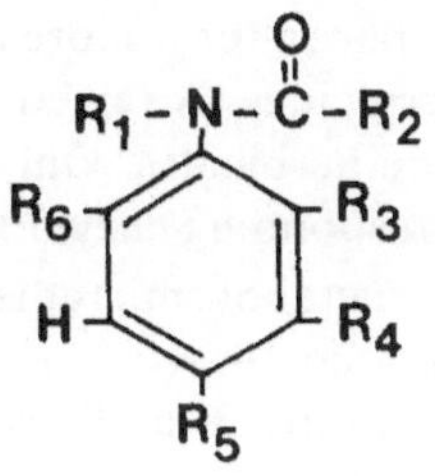

R_1 = an alkyl group or H
R_2 = alkyl group, CH_2Cl
R_3 = H, C_2H_5
R_4 = Cl, H
R_5 = CH_3, Cl, H
R_6 = H, C_2H_5, CH_3

FIGURE 9. General chemical structure of acylanilide herbicides.

1-methylurea], the acylanilide herbicides, monalide (4-chloro-alpha, alpha-dimethylvaleranilide) and propanil (3′,4′-dichloropropionanilide), the acylanilide fungicides, 2-chlorobenzanilide and 2,5-dimethyl-furan-3-carboxanilide, and the phenylcarbamate herbicide, propham, also induced the formation of the aryl acylamidase of *B. sphaericus*. However, the specific activities of the various extracts were considerably lower than that from crude extracts of linuron-induced cells. Production of the enzyme was initiated in the late exponential growth phase. A cell-free extract was prepared, and the enzyme was purified to homogeneity by ion-exchange chromatography, gel filtration, and polyacrylamide gel electrophoresis. The purification was 73-fold with a 2% yield. The enzyme was very labile during large-scale preparation and exhibited a pH optimum between 7.0 and 8.5. The enzyme lacked a demonstrable metal ion requirement, was quite sensitive to inhibition by sulfhydryl reagents and some metal ions, and lost 50% of its activity when held at 40°C for 9.5 min. The molecular weight of the enzyme was determined to be 75,000 daltons + or − 10%, and the specific activity of the purified linuron-induced enzyme was 130 nmol of substrate per minute per milligram protein.

It appears that the hydrolysis of the amide bond of several phenylurea herbicides ispossible by the action of acylamidase or arylacylamidase enzymes obtained from certain microorganisms. These enzymes are of particular interest for use in pesticide detoxification because they require no co-factors and exhibit relatively wide substrate specificity. Although the phenylurea herbicides have low mammalian toxicity (Table 1), the hydrolysis of these compounds may be desirable for destroying the phytotoxicity of the herbicides in question.

VI. ACYLANILIDES

A. Metabolism

The acylanilides are a group of herbicides that are effective weed killers because they provide selective toxicity at a low cost. They also exhibit relatively low mammalian toxicity. The basic structure of this class of herbicides is shown in Figure 9. The acylanilide herbicides resemble, structurally, the phenylurea and phenylcarbamate herbicides because all of these compounds contain a substituted aniline moiety. The chemistry and degradation of acylanilide herbicides was reviewed by Still and Herrett[59] and the reader is referred to this paper for a more detailed description concerning the metabolism of this group of herbicides. A review by Bartha and Pramer[97] also described the metabolism of acylanilide herbicides. Only the highlights of microbiological acylanilide degradation and corresponding enzymology are presented here.

One of the first microbial metabolism studies of the acylanilides was performed by Bartha,[98] in which the metabolic activities of soil microorganisms were investigated. Bartha concluded that 3,4-dichloroacylanilides were degraded microbiologically in soil and that the rate of degradation was inversely proportional to the length of the aliphatic side chain. The herbicides propanil, Dicryl [N-(3,4-dichlorophenyl)methylacrylamide,] Karsil [N-(3,4-dichlorophenyl)-2-methyl pentanamide], and propachlor (N-isopropyl-2-chloroacetanilide) were applied to soil at 550 ppm and their fate was investigated. The alkyl moieties of propanil, Dicryl, and Karsil were oxidized, liberating 3,4-dichloroaniline, which in turn could condense to 3,3′4,4′-tetrachloroazobenzene. Propachlor exhibited increased persistence in soil because of alkyl substitution of the acetanilide nitrogen.

Kaufman and Blake[99] studied the microbiological degradation of several herbicides containing aniline moieties. Among the herbicides examined, propachlor was found to be degraded by pure cultures of *A. ustus, F. oxysporum, F. solani, P. janthinellum* and *T. viride*. This herbicide could serve as a sole carbon and energy source and was metabolized by *F. oxysporum,* where it was observed that the microorganisms dehalogenated propachlor and formed hydroxypropachlor. Initial hydrolysis of the amide linkage of this compound was not necessary for degradation. The authors concluded that their results illustrated the adaptability and "omnivorous nature" of microbial populations from soil. Interestingly, all soil microorganisms studied demonstrated a different range of substrate metabolism.

The dissipation in soil of the alpha-chloroacetanilide herbicides alachlor, butachlor, and propachlor was investigated by Beestman and Deming.[100] They observed that the major avenue of dissipation of these compounds in soil was by microbial metabolism. The acetanilides were approximately 50 times more stable in sterilized soil than in nonsterilized soil and the investigators predicted no buildup of these herbicides when applied to soils at recommended application rates.

Smith and Phillips[101] reported the degradation of alachlor by *Rhizoctonia solani.* The organism was grown in media containing either 0, 1.2, 6.3, or 17.6 m*M* sucrose amended with either 0, 1, or 10 mg/ℓ technical grade alachlor. The media were inoculated with the fungus and shaken at 22°C. Alachlor did not influence sugar utilization or growth of the fungus, but fungal growth and alachlor degradation were proportional to the amount of sucrose in the media. Alachlor was not utilized as a sole source of carbon or energy, and growth and degradation were dependent upon the presence of a utilizable carbon source. No alachlor metabolites were described in this study.

In order to more closely examine the mechanisms of alachlor degradation, Tiedje and Hagedorn[102] used replacement cultures of the soil fungus, *Chaetomium globosum.* The fungus degraded alachlor and produced chloride ions and four identifiable metabolites, including 2-chloro-2′,6′-diethylacetanilide, 2′,6′-diethyl-N-(methoxymethyl)-aniline, 2,6-diethylaniline, and 1-chloroacetyl-2,3-dihydro-7-ethylindole. It is important to note that these investigators did not report the growth of the fungus at the expense of alachlor, but rather, simply the degradation of the herbicide. The authors suggested pathways of alachlor degradation and indicated that incubation of the fungus with 2-chloro-2′,6′-diethylacetanilide, 2,6-diethylaniline, and monochloroacetic acid resulted in further degradation of these alachlor metabolites. The authors maintained that the ability to degrade alachlor was not widely distributed among the soil fungi they examined.

The degradation of the herbicide butachlor (see Appendix) by several soil microorganisms was reported by Chen and Wu.[103] Replacement cultures grown at the expense of a carbon source other than butachlor were effective in degrading butachlor. Partial pathways of butachlor degradation have been proposed by these investigators. The main metabolic pathway involved debutoxymethylation of butachlor to form 2-chloro-2′,6′-diethylacetanilide. This compound was metabolized by either dehydrogenation cyclization to form N-chloroacetyl-7-ethyl-2,3-dihydroindole, or by dechlorination to produce 2-hydroxy-2′,6′-diethylacetanilide. Two other pathways were proposed and included either simultaneous debutoxymethoxylation and dechlorination to produce 2′,6′-diethylacetanilide or debutoxylation to form N-methyl-2-chloro-2′,6′-diethylacetanilide.

McGahen and Tiedje[104] reported the degradation of two acylanilide herbicides, Antor [2-chloro-N-(2′,6′-diethylphenyl)-N-methyl(ethylcarboxylate)acetamide] and Dual [2-chloro-N-(2′-ethyl-6′-methylphenyl)-N-(2-methoxy-1-methylethyl)acetamide] by resting cells of *C. globosum*. At least ten products were observed with Antor herbicide and at least eight products were observed with Dual. Generally, *C. globosum* removed one or both of the R groups from the nitrogen, dehydrogenated the ethyl substituent, and in some cases formed an indoline ring. The organisms also replaced the chloro, ethoxy, or methoxy substituents from R groups with a hydroxyl group.

B. Enzymology

Sharabi and Bordeleau[96] described organisms that were capable of degrading Karsil. Karsil could be degraded by whole cells or by cell-free extracts of a *Penicillium* sp. 2-methylvaleric acid and 3,4-dichloroaniline were produced. The Karsil acylamidase (EC 3.5.1) was induced by 0.1 m*M* Karsil and showed marked variation in enzyme activity for different substrates, including 22 various acylanilides (including the 3 herbicides Dicryl, propanil, and solan), 1 dimethylurea herbicide (diuron), and 1 phenylcarbamate herbicide (chlorpropham). After partial purification, the enzyme showed greatest activity for N-(phenyl)-butyramide, when it released 5.5 nmol of arylamine per minute per milligram. Activity of the enzyme generally increased with increased side-chain length up to four carbons. Substitution or branching of the N-acyl group adversely affected enzyme activity and a 2-chloro substitution on the side chain of propanil also reduced activity. Deacylation was influenced by the nature and position of ring substituents. A 2-chloro substitution on the ring prevented hydrolysis, possibly because of a steric effect.

As mentioned previously during the discussion of phenylurea herbicides, Kearney[83] reported that an enzyme responsible for degrading chlorpropham was isolated from *Pseudmonas striata* Chester. In this study, Kearney recognized that the phenyl carbamate-induced enzyme also showed specificity for 3,4-dichloroacetanilide and 3,4-dichloropionanilide.

Lanzilotta and Pramer[105] used propanil (3′,4′-dichloropropionanilide) with 0.16 g/ℓ of basal salts medium and isolated a strain of *Fusarium solani* from soil that could utilize the herbicide as a sole source of carbon and energy. The proposed pathway for the metabolism of propanil is shown in Figure 10. The primary metabolite, 3,4-dichloroaniline, accumulated in the medium up to a level of 80 μg/mℓ. This level of 3,4-dichloroaniline prohibited further utilization of the herbicide and allowed only minimal growth. When glucose (0.05%) and yeast extract (0.01%) were added to propanil-containing basal salts medium, the rate of propanil degradation increased, whereas reduced aeration retarded fungal growth and herbicide metabolism. The increased rate of propanil degradation due to the addition of glucose or yeast extract was assumed to be a result of increased cell population at the onset of propanil utilization. The investigators believed that the acylamidase involved probably functioned during usual metabolic activities of the fungus, acting on substrates more "normal" than propanil.

In further studies with the acylamidase of *F. solani*, Lanzilotta and Pramer[95] characterized the fungal aryl acylamidase (EC 3.5.1). Replacement cultures, previously grown on Czapek-Dox broth, were used to produce the enzyme when they were placed in 0.05 *M* phosphate buffer (pH 7.4) to which had been added either 150 ppm propanil or 270 ppm acetanilide. A crude enzyme extract was obtained from the acetanilide-containing medium and the enzyme was partially purified and concentrated by ammonium sulfate precipitation. The enzyme was active over a pH range of 7.5 to 9.0, was inactivated in 10 min at 50°C and showed maximum enzyme activity of 38°C. The enzyme exhibited a K_m of 0.195 m*M* for acetanilide. Propachlor was a competitive inhibitor of acetanilide hydrolysis. Lanzilotta and Pramer[95] also investigated the influence of chemical configuration on the rate of hydrolysis of various structurally related acylanilides. The effect of para-substitution generally decreased the rate of hydrolysis. Acetanilide was also hydrolyzed 42 times faster than propionanilide, and when a chlorine atom was present on the acyl chain of acetanilide, the rate of hydrolysis decreased approximately 67%. The phenylurea herbicides monuron and fenuron, and the phenylcarbamate herbicide, propham, were not hydrolyzed by this enzyme preparation.

Blake and Kaufman[106] also characterized acylanilide-hydrolyzing enzymes from *F. oxysporum* Schlecht. The enzymes were induced by propanil and PPG-124 (p-chlorophenylmethyl carbamate), and both enzymes were separated into fractions greater than 100,000 daltons or less than 50,000 daltons, and the greater than 100,000 daltons fraction exhibited 95% of the total amidase activity. The investigators were unable to demonstrate whether the amidases were similar, but reported that the two enzymes possibly were different.

The enzymes were produced by cells cultured in a basal salts medium containing 0.2% sucrose and 0.1% yeast extract. Cells were grown for 3 days at 25°C in the presence of either 50 ppm propanil or 10 ppm PPG-124, then harvested by filtration and immediately frozen and lyophilized. Active enzyme preparations were produced at 5°C by grinding the lyophilized cells and precipitating the proteins with ammonium sulfate. The proteins were then purified by ultrafiltration. Enzyme activities were determined by assaying the rate of chloroaniline formation. With the propanil-induced enzyme, the greatest specific activity of the enzyme was observed for the inducer, and was 9.25 nmol/min/mg protein. The PPG-124-induced enzyme expressed a specific activity of 66.66 nmol/min/mg protein.

The pH optimum of the greater than 100,000 dalton fraction of the propanil-induced enzyme was 10, whereas the pH optimum of the greater than 100,000 daltons fraction of the PPG-124-induced enzyme was 9. When the substrate specificities of both enzymes were investigated, it was noted that the acyl chain length definitely affected the rate of hydrolysis. The effect of number and position of chloro-substituents was also determined, and para-substitution appeared to increase the rate of hydrolysis, whereas ortho-substitution reduced the rate of hydrolysis. Both enzymes were capable of hydrolyzing acylanilides but were incapable of hydrolyzing the phenylurea herbicides linuron or diuron and the propanil-induced enzyme could not hydrolyze PPG-124 or chlorpropham. The apparent K_m of the propanil-induced enzyme was 2.3×10^{-5} *M*, whereas the apparent K_m of the PPG-124-induced enzyme was 1.0×10^{-3} *M*.

VII. PHENOXYACETATES

A. Metabolism

The phenoxyacetate herbicides, which include 2,4-dichlorophenoxyacetic acid (2,4-D), 2,4,5-trichlorophenoxyacetic acid (2,4,5-T), and 4-chloro-2-methylphenoxyacetic acid (MCPA) have been used extensively for the control of weeds since their introduction in the mid 1940s.[108] The basic structure of the phenoxyacetates is shown in Figure

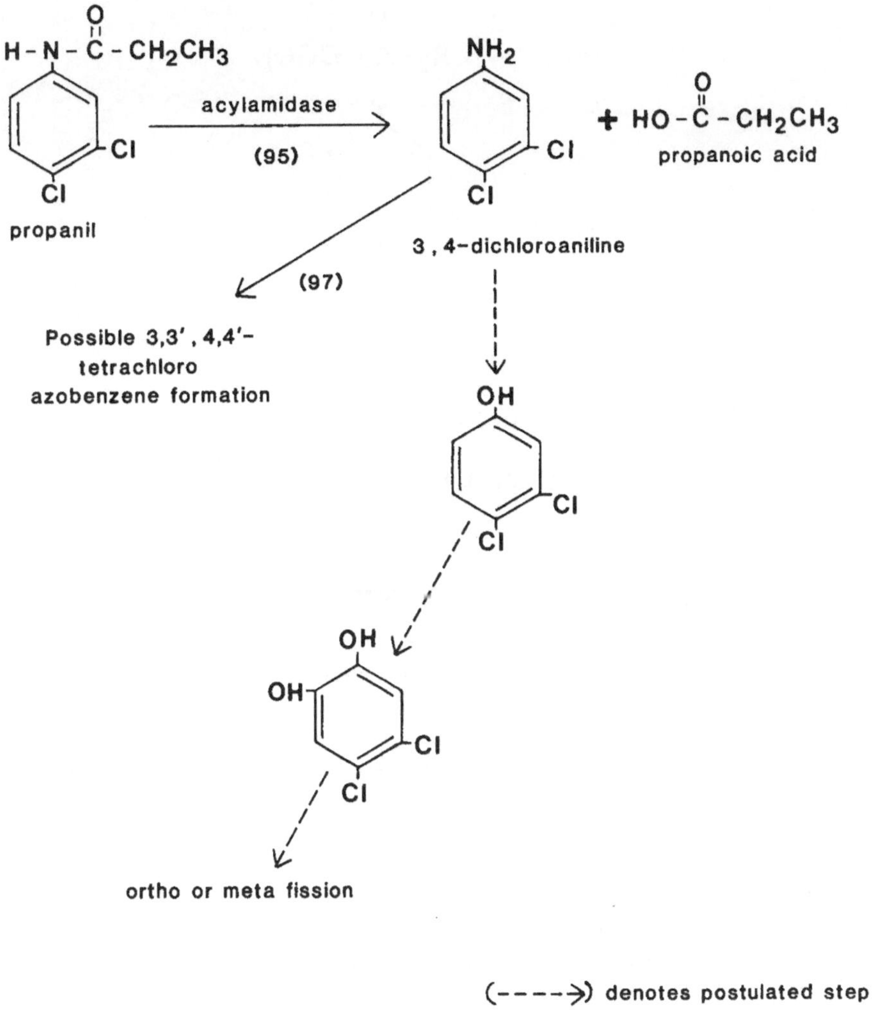

FIGURE 10. Proposed pathway for the metabolism of propanil.

11. Their selectivity allows for their use on croplands, lawns, and pastures.[108] The recent concern over the possible mutagenicity of phenoxyacetate herbicides and the dioxin (2,3,7,8-tetrachlorodibenzo-*p*-dioxin) contamination of 2,4,5-T has resulted in restrictions on their use and increased interest in their biochemistry and metabolism.

Loos has reviewed the chemistry, degradation, and mode of action of the phenoxyacetate herbicides,[108] and others have reviewed the microbial metabolism and degradation of these compounds in natural environments.[109,110] In general, 2,4-D disappears more rapidly than other phenoxyacetates after application to soil, usually within 2 to 3 weeks. MCPA and 2,4,5-T persist for a more variable period of time after application, ranging from 6 weeks or longer for MCPA and between 6 and 38 weeks for 2,4,5-T.[108] In aquatic environments, biodegradation of the phenoxyacetate herbicides occurs more slowly than in soil, which may be due to the lower cell densities of microorganisms encountered in these habitats.[111]

Many bacterial and fungal isolates have been obtained that can degrade phenoxyacetate herbicides. Among these are strains of *Pseudomonas, Achromobacter, Flavobacterium, Corynebacterium, Arthrobacter, Nocardia, Streptomyces,* and *Aspergillus.*[108] Of these organisms, the best characterized are those that are involved in the degrada-

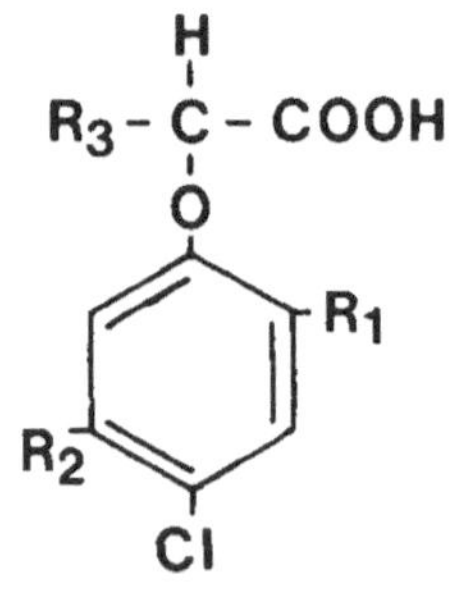

$R_1 = Cl$ or CH_3

$R_2 = H$ or Cl

$R_3 = H$ or CH_3

FIGURE 11. General chemical structure of phenoxyacetate herbicides.

tion of 2,4-D. A strain of *Arthrobacter* isolated by Loos et al.[112] was especially well studied. This organism was isolated from silt loam and rapidly oxidized 2,4-D, 2- and 4-chlorophenoxyacetate, 3,5-di- and 4-chlorocatechol, catechol, 2,4-dichlorophenol, and 2- and 4-chlorophenol when grown on 2,4-D. But when cells of the *Arthrobacter* sp. were grown on citrate, these compounds were oxidized slowly or not at all, indicating that the enzyme system mediating the degradation of 2,4-D was inducible. The first intermediate in the degradation pathway of 2,4-D was 2,4-dichlorophenol, and the aromatic ring of 2,4-D was eventually cleaved with all chlorine converted to free chloride.[113]

Tiedje et al.[107] outlined the entire degradation sequence for 2,4-D in the *Arthrobacter* isolate. 2,4-D was first converted to 2,4-dichlorophenol, which was subsequently converted to 3,4-dichlorocatechol, *cis,cis*-2,4-dichloromuconic acid, 2-chloro-4-carboxy-methylene-but-2-enolide, chloromaleylacetic acid, and succinate (Figure 12). An ortho-fission mechanism was involved in ring cleavage, which resulted in the formation of the dichloromuconic acid.

In a related investigation, Tiedje and Alexander[114] determined that the acetate moiety released when the ether linkage of 2,4-D was cleaved by the *Arthrobacter* sp. was metabolized to alanine and a volatile compound. It was proposed that glyoxylate was the initial product formed upon the removal of the acetate side chain from 2,4-D, and that alanine was produced as the result of a condensation reaction.

Evans et al.[115] isolated pseudomonad strains from soil that could utilize 2,4-D as a sole carbon source. These organisms possessed an ortho-fission mechanism similar to that reported for the *Arthrobacter* sp., and the primary degradation pathway for these isolates contained the same intermediates as the *Arthrobacter* sp. degradation pathway.

Kilpi et al.[116] isolated a species of *Pseudomonas* from a field soil treated with the herbicide mecoprop, (±)-2-methyl-4-chlorophenoxypropionic acid (MCPP), that could utilize either 2,4-D, MCPA, benzoic acid, or salicylic acid as a sole carbon source. When the cells were grown on 2,4-D or MCPA, an oxygenase (catechol-1,2-dioxygenase, EC 1.13.11.1) was induced that catalyzed the degradation of 3-methyl-5-chlorocatechol, 3,5-dichlorocatechol, and 3-chlorocatechol by means of an ortho-fission mechanism. The enzyme preparation was especially active toward 3-methyl-5-chlorocatechol, which is an intermediate in the degradation of MCPA.

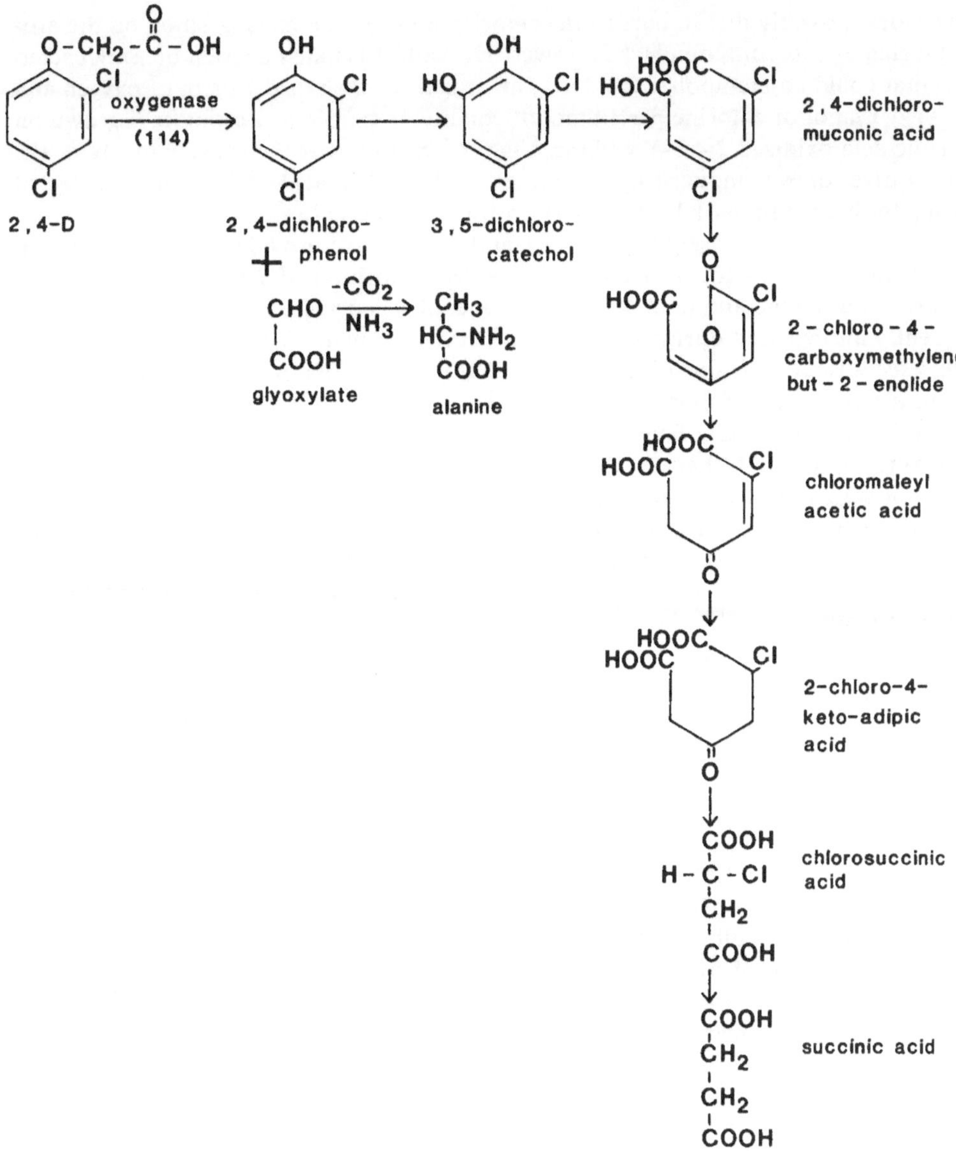

FIGURE 12. Pathway for the metabolism of 2,4-D.

Another phenoxy herbicide, 2-(2,4,5-trichlorophenoxy) propionic acid (silvex) was found by Ou and Sikka[117] to be metabolized by a mixed culture of a *Pseudomonas* sp. and an *Achromobacter* sp. isolated from a farm pond water. When the two isolates were separated, neither could metabolize silvex. Together, they could not utilize silvex as a sole carbon source, but could degrade it in the presence of another carbon source such as yeast extract. The degradation process appeared to be co-metabolic and synergistic. Evidence for ring cleavage was obtained when radiolabeled carbon dioxide was evolved from the mixed culture when it was incubated with ^{14}C-ring-labeled silvex. Resting cell suspensions accumulated 2,4,5-trichlorophenol, and cells grown with silvex were capable of rapidly degrading 2,4,5-trichlorophenol and 3,5-dichlorocatechol without a lag, indicating that these two compounds were intermediates in the silvex degradation pathway.

The herbicide 2,4,5-T is more resistant to degradation than other phenoxyacetate herbicides, possibly due in part to the chlorine atom in the meta position on the aromatic ring of this compound.[118] However, Horvath[119] isolated a strain of *Brevibacterium* that could co-metabolize 2,4,5-T. The organism consumed 1 μmol of oxygen and released 1 μmol of chlorine per 1 μmol of compound oxidized. Resting cells grown on benzoic acid oxidized 2,4,5-T without a lag, suggesting that the enzyme involved was constitutive, or was induced by the presence of benzoic acid. A metabolite detected during the degradation of 2,4,5-T was identified as 3,5-dichlorocatechol.

Recent reports by Rosenberg and Alexander[120,121] have provided more information concerning the breakdown of 2,4,5-T. The decomposition of 2,4,5-T in soil was examined, and the metabolite 2,4,5-trichlorophenol was detected. This compound subsequently disappeared during the course of the experiment. A strain of *Pseudomonas fluorescens* isolated from soil degraded in 80 hr about 70% of the 2,4,5-T added (20 μg) to a basal salts medium containing glucose, with 60% of the herbicide recovered as 2,4,5-trichlorophenol. This culture did not use 2,4,5-T as a sole source of carbon, however, and did not liberate radiolabeled carbon dioxide from ring-labeled 2,4,5-T. In a soil suspension, 2,4,5-trichlorophenol was converted by microorganisms to 3,5-dichlorocatechol, 4-chlorocatechol, and succinate, which were detected by gas chromatography and mass spectrometry and also *cis,cis*-2,4-dichloromuconate,2-chloro-4-carboxymethylene-but-2-enolide, and chlorosuccinate, which were detected by gas chromatography and thin layer chromatography. From soil and sewage, 52 bacterial cultures were isolated and 41 of these metabolized 2,4-D. Nineteen cultures could metabolize 2,4-D and 2,4,5-T, but none grew using the herbicides as carbon sources. Cometabolism was responsible for the formation of phenolic intermediates, aromatic ring cleavage, and liberation of chloride.

In 1977, Pemberton and Fisher[122] reported that a class of plasmids carried genetic information responsible for the degradation of 2,4-D and postulated that the presence of these plasmids which coded for pesticide-degrading enzymes contributed greatly to the disappearance of 2,4-D in soil. Further work by Fisher et al.[123] demonstrated that the rapid degradation of 2,4-D and MCPA by a strain of *Alcaligenes paradoxus* was mediated by a 58M-dalton conjugal plasmid. No other phenotypic characteristics were shown to be encoded by this plasmid. The enzyme controlling the first step in the degradation pathway was believed to be specific for 2,4-D and MCPA, since phenoxyacetic acid was not metabolized.

B. Enzymology

A cell-free extract obtained by Loos et al.[124] from the previously discussed *Arthrobacter* sp. was capable of catalyzing the degradation and dehalogenation of 2,4-D, MCPA, and 4-chlorophenoxyacetic acid. First, the side chain of 2,4-D, attached by ether linkage to the aromatic ring, was detached without prior hydroxylation of the ring to form 2,4-dichlorophenol. In 3 hr, 92% of the organic chlorine in 2,4-D was released. This information can be used to calculate a specific activity of 0.3 nmol/min/mg protein for the overall dehalogenation reactions. Data reported by Tiedje and Alexander[114] can be used to estimate a specific activity of 0.1 nmol/min/mg protein for their enzyme preparation [EC 1.14.99 (monooxygenase) or possibly EC 3.3.2 (hydrolase)] when it catalyzed the conversion of 2,4-D to 2,4-dichlorophenol (Table 2).

Bollag et al.[125] continued this work and found that enzymes from the *Arthrobacter* sp. hydroxylated 2,4-dichlorophenol to form 3,5-dichlorocatechol. From their data a specific activity of 6.2 nmol/min/mg protein can be calculated. NADPH and oxygen were required for the reaction, indicating the involvement of a mixed function oxidase (EC 1.14.13.7). The 3,5-dichlorocatechol was subsequently converted to an ultraviolet-absorbing compound (2-chloro-4-carboxymethylene-but-2-enolide) and free chloride.[126] Oxygen was required for this reaction, but no pyridine nucleotides or iron were needed. The specific activity of the enzyme preparation for this reaction can be estimated as 4.4 nmol/min/mg protein. The lyophilized, cell-free enzyme extract retained its activity toward chlorocatechol for several months.

Bollag et al.[127] obtained an isolate of *Flavobacterium peregrinum* from a soil enrichment which was capable of degrading MCPA. The enzymatic activity of this organism and the *Arthrobacter* sp. isolated by Loos et al. toward MCPA was then compared. Both cell-free extracts catalyzed the conversion of MCPA to 4-chloro-2-methylphenol. The specific enzyme activity for dehalogenation can be estimated to be 0.1 nmol/min/mg protein for the *Arthrobacter* extract (Table 2) and 0.01 nmol/min/mg protein for the extract from the *Flavobacterium* sp. The extract from the *Arthrobacter* sp. also catalyzed the specific dehalogenations of 4-chloro-2-methylphenol and 4-chlorocatechol. The *Flavobacterium* sp. extract did not release chloride from the phenol. Subsequently, the *Arthrobacter* extract was fractionated by gel filtration. Active fractions were not capable of hydroxylating 4-chloro-2-methylphenol to a catechol unless NADPH was added. Dechlorination occurred only after the catechol was formed.

VIII. DISCUSSION

Several enzymes capable of degrading a wide variety of pesticides have been described in this chapter. None of these enzymes is practical for immediate routine industrial use, but the potential importance of their development for pesticide disposal has been indicated. Enzyme-mediated pesticide disposal can be applied to most situations where removal of a hazardous chemical is desired. In agricultural practices, methods for detoxifying pesticides are nonexistent, yet unwanted pesticides are often generated by miscalculations in mixing or spraying, or by residues remaining in containers. Soil spills are not uncommon on farms either, and safe methods for detoxifying spilled pesticides are needed. In any of these situations a suitable pesticide detoxifying enzyme preparation could be used by the consumer to allow safe and efficient disposal of unwanted chemicals. The enzymatic removal of parathion from pesticide containers has already been demonstrated[10] when it was observed that 85 g of residual parathion in a 200-ℓ metal drum were hydrolyzed after parathion hydrolase was added. Within 16 hr, 90% of the parathion present in the drums as a 48% emulsifiable concentrate was hydrolyzed. Methyl parathion, diazinon, and dursban at concentrations up to 1% technical compound could also be enzymatically hydrolyzed.

The feasibility of using enzyme preparations to clean up soil spills has also been demonstrated. Recent studies by Domsch et al.[128] and Munnecke and Barik[129] indicated that when parathion hydrolase was added to soils, it efficiently hydrolyzed 1% parathion or diazinon within 24 hr to less toxic and unstable products.

Enzymes capable of degrading pesticides could also be used in agriculture to prevent carry-over of crop protection chemicals. Spray tanks and other equipment containing phenoxyacetate, phenylurea, acylanilide, or other herbicides could be treated with an appropriate enzyme to remove residual pesticide. This would prevent carry-over and accidental phytotoxicity during subsequent spraying procedures.

Detoxification by enzymes rather than by whole microbial cells is particularly beneficial because enzymes sometimes can tolerate environmental extremes better than whole microbial cells. Extremes of pH and temperature, as well as high salt and solvent concentrations are often encountered in pesticide production wastewaters. The enzyme parathion hydrolase, for example, could tolerate salt concentrations up to 10%, solvent concentrations of 1%, and temperatures up to 50°C. These three parameters would have prevented growth of the *Pseudomonas* sp. that produced the hydrolase.[10] The use of enzymes is also attractive because the transport of pesticides into whole microbial cells can be problematic. Such membrane transport problems could be avoided when soluble enzymes are employed in disposal processes.

The use of immobilized enzymes in a continuous flow column for detoxification of pesticide production wastewaters has also been described.[11] This system was examined to determine its ability to hydrolyze residual parathion in process wastewaters before discharge into waste treatment systems. The enzyme column was effective in reducing parathion levels from 10 ppm to 500 ppb in approximately 1 min and the enzyme remained active under continuous use conditions for over 70 days.

Enzymes that have the greatest importance for pesticide detoxification are those that can function without co-factors or co-enzymes and can detoxify a pesticide molecule by a hydrolytic or other simple enzymatic reaction. Presently, there are some enzymes that would be useful for detoxification reactions but are not likely candidates for immediate development because they require co-factors or co-enzymes. For example, the oxygenase enzymes (EC 1.14.12 and EC 1.14.13) require molecular oxygen and either NADPH or NADH to initiate aromatic ring fission. The development of these enzymes for large-scale industrial applications will be prohibited until economical methods for co-factor regeneration are discovered. Carboxy-lyase enzymes (EC 4.1.1) could also be important degradative enzymes for industrial pesticide detoxification use, but these require thiamin-diphosphate or pyridoxal-phosphate, and until an inexpensive source of these co-factors is available, carboxy-lyase enzymes cannot be developed for wastewater treatment.

Some limitations of pesticide detoxification by enzyme technology should be considered here. It is a fact that some compounds are not detoxified by enzymatic hydrolysis, but instead, the metabolites have increased toxicity (Table 1). In such instances, initial enzymatic treatment of a parent compound may cause the molecule to subsequently be degraded more rapidly by whole microbial cells. Some pesticides do not pose toxicity problems, but because they are used in such large quantities there is a need for proper detoxification technology. Because enzyme mediated disposal can occur under mild conditions, we feel that the enzymatic treatment of low toxicity compounds is highly desirable.

IX. FUTURE DIRECTIONS AND GOALS

The demand for industrial enzymes will average an 8% growth rate over the 6-year period between 1979 and 1985.[130] Although the use of enzymes for detoxifying pesticides has been limited on the industrial scale, we feel that large-scale use is quite feasible because such a wide variety of pesticide detoxifying enzymes have already been described. There are other enzymes that have not yet been characterized in depth that may also serve to detoxify pesticides. One such group of enzymes is the halidohydrolases. It would be desirable if enzyme systems could be developed to dehalogenate aromatic hydrocarbons since the basic nucleus of many pesticides is a halogenated aromatic ring. Two halidohydrolases have been described that were induced by chloroacetate and dichloroacetate. These enzymes used a hydroxyl group from water to replace the halogen in the substrate.[131] Dechlorination of aromatic compounds has been described by Janke and Fritsche.[132]

In our laboratories, we are searching for new enzymes that can detoxify organophosphate, dithioate, acylanilide, and other pesticides, including the herbicide glyphosphate [N-(phosphonomethyl)glycine]. We feel that enzymes capable of hydrolyzing these pesticides can be developed and ultimately considered for large-scale industrial use. We are particularly interested in developing enzymes for detoxifying pesticide production wastewaters. Once these detoxification enzymes have been developed, the search for and development of other new enzymes, including those that dehalogenate or those that require co-factors, may be pursued.

X. APPENDIX

Common name	Chemical name
Acephate	O-S-Dimethyl acetylphosphoramidothioate
Alachlor	2-Chloro-2′,6′-diethyl-N-(methoxymethyl) acetanilide
Aldicarb	2-Methyl-2-(methylthio) propionaldehyde O-(methylcarbomoyl)oxime
Antor	N-Chloroacetyl-N-(2,6-diethylphenyl)glycine ethyl ester
Aspon	1,2,4,5,6,7,8,8-Octachlor-2,3,3,3a,4,7,7a-hexahydro-4,7-methanoindane
Azinphos-methyl	O,O-Diethyl-s-(4-oxo-3-H-1,2,3-benzotriazine-3yl)-methyl-dithiophosphate
Butachlor	N-(Butoxymethyl)-2-chloro 2′,6′-diethylacetanilide
Carbaryl	1-Naphthyl methylcarbamate
Carbofuran	2,3-Dihydro-2,2-dimethyl-7-benzofuranyl-N-methylcarbamate
Carboxin	5,6-Dihydro-2-methyl-1,4-oxathiin-3-carboxanilide
Carbophenothion	S-[(*p*-Chlorophenylthio)methyl] O-diethyl phosphorodithioate
Chlorbromouron	3-(4-Bromo-3-chlorophenyl)-1-methoxy-1-methylurea
Chlormephos	S-Chlormethyl-O,O-diethyl phosphorothiolothionate
Chlorothion	O,O-Dimethyl-O-(3-chloro-4-nitrophenyl) carbamate
Chloroxuron	3-[p-(p-Chlorophenoxy)phenyl]-1,1-dimethylurea
Chlorpropham	Isopropyl-N-3-chlorophenyl-carbamate
Cyanophos	O-(4-Cyanophenyl)O,O-dimethyl phosphorothioate
2,4-D	(2,4-Dichlorophenoxy)acetic acid
DFP	Di-isopropylfluorophosphate
Diallate	S-(2,3-Dichloroallyl)-diisopropyl-thiocarbamate
Diazinon	O,O-Diethyl O-(2-isopropyl-4-methyl-6-pyrimidinyl) phosphorothioate
Dichlorvos	2,2-Dichlorovinyl O,O-dimethyl phosphate
Dicryl	3′,4′-Dichloro-2-methylacrylanilide
Dimethoate	O,O-Dimethyl S-(N-methylcarbamoylmethyl) phosphorodithioate
Disulfoton	O,O-Diethyl S-[2-(ethylthio)ethyl] phosphorodithioate
Diuron	3-(3,4-Dichlorophenyl)-1,1-dimethylurea
Dursban	O,O-Diethyl O-(3,5,6-trichloro-2-pyridyl) phosphorothioate
EPN	O-Ethyl O-(4-nitrophenyl) phenyl phosphonothioate
EPTC	S-Ethyl dipropylthiocarbamate
Ethion	O,O,O,O-Tetraethyl-S,S-methylene bisphosphorodithioate
Fenitrothion	O,O-Dimethyl O-(3-methyl-4-nitrophenyl) phosphorothioate
Fensulfothion	O,O-Diethyl-O-[4-methylsulfinyl)phenyl] phosphorothioate
Fenthion	O,O-Dimethyl-O-[3-methyl-4-(methylthio)phenyl] phosphorothioate
Fenuron	3-Phenyl-1,1-dimethylurea
Fluometuron	1,1-Dimethyl-3-(a,a,a-trifluoro-m-tolyl) urea
Karsil	N-(3,4-Dichlorophenyl)-2-methyl pentanamide
Linuron	3-(3,4-Dichlorophenyl)-1-methoxy-1-methylurea
Malathion	O,O-Dimethyl-S (1,2-dicarbethoxyethyl)-phosphorodithioate
MCPA	4-Chloro-2-methyl phenoxyacetic acid
Mecoprop	2-(4-Chloro-2-methylphenoxy) propionic acid
Methyl parathion	O,O-Dimethyl-O-*p*-nitrophenyl-phosphorothioate
Metobromuron	3-(p-Bromophenyl)-1-methoxy-1-methylurea

Metolachlor	2-Chloro-N-(2-ethyl-6-methylphenyl)-N-(2-methoxy-1-methylethyl) acetamide
Mexacarbate	4-Dimethylamino-3,5-xylyl N-methylcarbamate
Monalide	N-(4-Chlorophenyl)-a,a-dimethylvaleramide
Monocrotophos	(E)-Dimethyl 1-methyl-3-(methylamino)-3-oxo-1-propenyl phosphate
Monolinuron	3-(4-Chlorophenyl)-1-methoxy-1-methylurea
Monuron	3-(p-Chlorophenyl)-1,1-dimethylurea
Neburon	1-N-Butyl-3-(3,4-dichlorophenyl)-1-methylurea
Paraoxon	O,O-Diethyl-O-p-nitrophenylphosphate
Parathion	O,O-Diethyl O-p-nitrophenyl phosphorothioate
PPG-124	p-Chlorophenyl methyl carbamate
Propachlor	2-Chloro-N-(1-methylethyl)-N-phenylacetamide
Propanil	N-(3,4-Dichlorophenyl) propanamide
Propetamphos	(E)-O-2-Isopropoxycarbonyl-1-methylvinyl-O-methyl ethylphosphoramidothioate
Propham	Isopropyl-N-phenylcarbamate
Pyracarbolid	2-Methyl-5,6-dihydro-4-H-pyran-3-carboxylic acid anilide
Quinalphos	O,O-Diethyl-O-2-quinoxalinyl phosphorothioate
Siduron	1-(2-Methylcyclohexyl)-3-phenylurea
Silvex	2-(2,4,5-Trichlorophenoxy)propionic acid
Solan	3′-Chloro-2-methyl-p-valerotoluidide
2,4,5-T	2,4,5-Trichlorophenoxyacetic acid
TEPP	Tetraethyl pyrophosphate
Triallate	S-(2,3,3-Trichloroallyl-diisopropyl-thiocarbamate)
Triazophos	1-Phenyl-1,2,4-triazolyl-3-(O,O-diethylthionophosphate)
Trichlorfon	Dimethyl (2,2,2-trichloro-1-hydroxyethyl) phosphonate

REFERENCES

1. **Alexander, M.,** Biodegradation: problems of molecular recalcitrance and microbial infallibility, *Adv. Appl. Microbiol.,* 7, 35, 1965.
2. **Horvath, R. S.,** Microbial co-metabolism and the degradation of organic compounds in nature, *Bacteriol. Rev.,* 36, 146, 1972.
3. **Storck, W. J.,** Pesticide profits belie mature market status, *Chem. Eng. News,* 58, 10, 1980.
4. **Murry, C.,** Senate panel probes kepone disaster, *Chem. Eng. News,* 54, 17, 1976.
5. **Munnecke, D. M., Day, H. R., and Trask, H. W.,** Review of pesticide disposal research, No. SW 527, U.S. *EPA Publication,* Cincinnati, 1976, 76.
6. **Munnecke, D. M.,** Chemical, physical, and biological methods for the disposal and detoxification of pesticides, *Residue Rev.,* 70, 1, 1979.
7. **Scurlock, A. C., Lindsey, A. W., Fields, T., and Huber, D. R.,** Incineration in hazardous waste management, EPA Report, OSWMP, EPA/530/SW-141, Environmental Protection Agency, Washington, D.C., 1975, 104.
8. **Bouwer, H.,** Use of the earth's crust for treatment or storage of sewage effluent and other waste liquids, *CRC Crit. Rev. Environ. Contr.,* 111, 1976.
9. **Ottinger, R. S. and Blumenthal, J. L.,** Recommended methods of reduction, neutralization, recovery or disposal of hazardous waste. V. Pesticides and cyanides, EPA Report NTIS, PB-224-579, Environmental Protection Agency, Springfield, Va., 1974, 144.
10. **Munnecke, D. M.,** Enzymatic detoxification of waste organophosphate pesticides, *J. Agric. Food Chem.,* 28, 105, 1980.
11. **Munnecke, D. M.,** Hydrolysis of organophosphate insecticides by an immobolized-enzyme system, *Biotechnol. Bioeng.,* 21, 2247, 1979.

12. **Christensen, H. E.**, Registry of toxic effects of chemical substances, U.S. Department Health, Education and Welfare, National Institute for Occupational Safety and Health, Rockville, Maryland, 1976, 1245.
13. **Adamson, J. and Inch, T. D.**, Possible relationships between structure and mechanism of degradation of organophosphorus insecticides in the soil environment, *Proc. 7th Br. Insecticide Fungicide Conf.*, 1, 15, 1973.
14. **Paris, D. F. and Lewis, D. L.**, Chemical and microbial degradation of ten selected pesticides in aquatic systems, *Residue Rev.*, 45, 95, 1973.
15. **Kaufman, D. D.**, Degradation of pesticides by soil microorganisms, in *Pesticides in Soil and Water*, Guenzi, W. D., Ed., Soil Science Society of America, Madison, Wis., 1974, 133.
16. **Tu, C. M. and Miles, J. R. W.**, Interactions between pesticides and soil microbes, *Residue Rev.*, 64, 17, 1976.
17. **Lavegillia, J. and Dahm, P. A.**, Degradation of organophosphorus and carbamate insecticides in the soil and by microorganisms, *Annu. Rev. Entomol.*, 22, 483, 1977.
18. **Sethunathan, N., Siddaramappa, R., Rajaram, K. P., Barik, S., and Wahid, P. A.**, Parathion: residues in soil and water, *Residue Rev.*, 68, 91, 1977.
19. **Lichtenstein, E. P., Fuhremann, T. W., and Schulz, K. R.**, Effect of sterilizing agents in persistence of parathion and diazinon in soil and water, *J. Agric. Food Chem.*, 16, 870, 1968.
20. **Barik, S. and Sethunathan, N.**, Biological hydrolysis of parathion in natural ecosystems, *J. Environ. Qual.*, 7, 346, 1978.
21. **Getzin, L. W. and Rosefield, I.**, Organophosphorus insecticides degradation by heat-labile substances in soil, *J. Agric. Food Chem*,. 16, 598, 1968.
22. **Munnecke, D. M. and Hsieh, D. P. H.**, Pathways of microbial metabolism of parathion, *Appl. Environ. Microbiol.*, 31, 63, 1976.
23. **Katan, J., Fuhremann, T. W., and Lichtenstein, E. P.**, Binding of 14C-parathion in soil: a reassessment of pesticide persistence, *Science*, 193, 891, 1976.
24. **Katan, J. and Lichtenstein, E. P.**, Mechanism of production of soil bound residues of 14C-parathion by microorganisms, *J. Agric. Food Chem.*, 25, 1404, 1977.
25. **Sethunathan, N.**, Degradation of parathion in flooded acid soils, *J. Agric. Food Chem.*, 21, 602, 1973.
26. **Siddaramappa, R., Rajaram, K. P., and Sethunathan, N.**, Degradation of parathion by bacteria isolated from flooded soil, *Appl. Microbiol.*, 26, 846, 1973.
27. **Ferris, I. G. and Lichtenstein, E. P.**, Interactions between agricultural chemicals and soil microflora and their effects on the degradation of 14C-parathion in a cranberry soil, *J. Agric. Food Chem.*, 28, 1011, 1980.
28. **Barik, S., Wahid, P. A., Ramakrishna, C., and Sethunathan, N.**, A change in the degradation pathway of parathion after repeated applications to flooded soil, *J. Agric. Food Chem.*, 27, 1391, 1979.
29. **Hsieh, D. P. H. and Munnecke, D. M.**, Accelerated microbial degradation of concentrated parathion, in *Fermentation Technology Today*, Terui, G., Ed., Society of Fermentation Technology, Tokyo, Japan, 1972, 551.
30. **Munnecke, D. M. and Hsieh, D. P. H.**, Microbial decontamination of parathion and *p*-nitrophenol in aqueous media, *Appl. Microbiol.*, 28, 212, 1974.
31. **Daughton, C. G. and Hsieh, D. P. H.**, Parathion utilization by microbial symbionts in a chemostat, *Appl. Environ. Microbiol.*, 34, 175, 1977.
32. **Eto, M.**, Organophosphorus pesticides, in *Organic and Biological Chemistry*, CRC Press, Boca Raton, Fla., 1974, 386.
33. **Mounter, L. A., Baxter, R. F., and Chanutin, A.**, Dialkylfluorophosphatases of microorganisms, *J. Biol. Chem.*, 215, 699, 1955.
34. **Mounter, L. A. and Tuck, K. D.**, Dialkyl-fluorophosphatases of microorganisms. II. Substrate specificity studies, *J. Biol. Chem.*, 221, 537, 1956.
35. **Zech, R. and Wigand, K. D.**, Organophosphate-detoxifying enzymes in E. coli. Gelfiltration and isoelectric focussing of DFPase, paraoxonase and unspecific phosphohydrolases, *Experientia*, 15, 157, 1975.
36. **Sethunathan, N. and Yoshida, T.**, A *Flavobacterium* sp. that degrades diazinon and parathion, *Can. J. Microbiol.*, 19, 873, 1973.
37. **Brown, K. A.**, Phosphotriesterases of *Flavobacterium* sp., *Soil. Biol. Biochem.*, 12, 105, 1980.
38. **Adhya, T. K., Barik, S., and Sethunathan, N.**, Hydrolysis of selected organophosphorus insecticides by two bacteria isolated from flooded soil, *J. Appl. Bacteriol.*, 50, 1981, in press.
39. **Munnecke, D. M.**, Enzymatic hydrolysis of organophosphate insecticides, a possible pesticide disposal method, *Appl. Environ. Microbiol.*, 32, 7, 1976.
40. **Munnecke, D. M.**, unpublished data, 1980.
41. **Rosenberg, A. and Alexander, M.**, Microbial cleavage of various organophosphorus insecticides, *Appl. Environ. Microbiol.*, 37, 886, 1979.

42. **Barik, S. and Sethunathan, N.**, Metabolism of nitrophenols in flooded soils, *J. Environ. Qual.*, 7, 349, 1978.
43. **Barik, S., Siddaramappa, R., and Sethunathan, N.**, Metabolism of nitrophenols by bacteria isolated from parathion-amended soil, *Antonie van Leeuwenhoek; J. Microbiol. Serol.* 42, 461, 1976.
44. **Gundersen, K. and Jensen, H. L.**, A soil bacterium decomposing organic nitro-compounds, *Acta Agric. Scand.*, 6, 100, 1956.
45. **Jensen, H. L. and Lautrup-Larsen, G.**, Microorganisms that decompose nitroaromatic compounds, with special reference to dinitro-orthocresol, *Acta Agric. Scand.*, 17, 115, 1967.
46. Degradation of Synthetic Organic Molecules in the Biosphere — Natural, Pesticidal and Various Other Man-Made Compounds, NAS, Proc. of a Conf. San Francisco, California, June 12—13, 1971, National Academy of Sciences, Washington, D.C., 1972, 350.
47. **Konrad, J. G., Chesters, G., and Armstrong, D. E.**, Soil degradation of malathion, a phosphorodithioate insecticide, *Soil Sci. Soc. Am. Proc.*, 33, 259, 1969.
48. **Walker, W. W. and Stojanovic, B. J.**, Microbial versus chemical degradation of malathion in soil, *J. Environ. Qual.*, 2, 229, 1973.
49. **Bourquin, A. W.**, Microbial-malathion interaction in artificial salt-marsh ecosystems — effect and degradation, EPA, *Ecol. Res. Ser.*, 1975, 41.
50. **Bourquin, A. W.**, Degradation of malathion by salt-marsh microorganisms, *Appl. Environ. Microbiol.*, 33, 356, 1977.
51. **Mostafa, I. Y., Bahig, M. R. E., Fakhr, I. M. I., and Adam, Y.**, Metabolism of organophosphorus insecticides. XIV. Malathion breakdown by soil fungi, *Z. Naturforsch.*, 27b, 1115, 1972.
52. **Walker, W. W.**, Chemical and microbiological degradation of malathion and parathion in an estuarine environment, *J. Environ. Qual.*, 5, 210, 1976.
53. **Paris, D. F., Lewis, D. L., and Wolfe, N. Lee**, Rates of degradation of malathion by bacteria isolated from aquatic system, *Environ. Sci. Technol.*, 9, 135, 1975.
54. **Lewis, D. L., Paris, D. F., and Baughman, G. L.**, Transformation of malathion by a fungus, *Aspergillus oryzae*, isolated from a freshwater pond, *Bull. Environ. Contam. Toxicol.*, 13, 596, 1975.
55. **Mostafa, I. Y., Fakhr, I. M. I., Bahig, M. R. E., and El-Zawahry, Y. A.**, Metabolism of organophosphorus insecticides. XIII. Degradation of malathion by *Rhizobium* sp., *Arch. Mikrobiol.*, 86, 221, 1972.
56. **Matsumura, F. and Boush, G. M.**, Malathion degradation by *Trichoderma viride* and a *Pseudomonas* species, *Science*, 153, 1278, 1966.
57. **Getzin, L. W. and Rosefield, I.**, Partial purification and properties of a soil enzyme that degrades the insecticide malathion, *Biochem. Biophys. Acta*, 235, 442, 1971.
58. **Barik, S. and Munnecke, D. M.**, unpublished data, 1980.
59. **Still, G. G. and Herrett, R. A.**, Methylcarbamates, carbanilates, and acylanilides, in *Herbicides — Chemistry, Degradation, and Mode of Action*, Vol. 2, 2nd ed., Kearney, P. C. and Kaufman, D. D., Eds., Marcel Dekker, New York, 1976, 609.
60. **Ryan, A. J.**, The metabolism of pesticidal carbamates, *CRC Crit. Rev. Toxicol.*, p. 33, 1971.
61. **Coppedge, J. R., Bull, D. L., and Ridgway, R. L.**, Movement and persistence of aldicarb in certain soils, *Arch. Environ. Contam. Toxicol.*, 5, 129, 1977.
62. **Andrawes, N. R., Bagley, W. P., and Herrett, R. A.**, Fate and carryover properties of Temik aldicarb pesticide (2-methyl-2-(methylthio) propionaldehyde *O*-(methyl-carbamoyl) oxime) in soil, *J. Agric. Food Chem.*, 19, 727, 1971.
63. **Richey, F. A., Jr., Bartley, W. J., and Sheets, K. P.**, Laboratory studies on the degradation of (the pesticide) aldicarb in soils, *J. Agric. Food Chem.*, 25, 47, 1977.
64. **Jones, A. S.**, Metabolism of aldicarb by five soil fungi, *J. Agric. Food Chem.*, 24, 115, 1976.
65. **Rodriguez, L. D. and Dorough, H. W.**, Degradation of carbaryl by soil microorganisms, *Arch. Environ. Contam. Toxicol.*, 6, 47, 1977.
66. **Liu, S.Y. and Bollag, J. M.**, Metabolism of carbaryl by a soil fungus, *J. Agric. Food Chem.*, 19, 487, 1971.
67. **Bollag, J. M. and Liu, S. Y.**, Hydroxylations of carbaryl by soil fungi, *Nature (London)*, 236, 177, 1972.
68. **Sikka, H. C., Miyazaki, S., and Lynch, R. S.**, Degradation of carbaryl and 1-naphthol by marine microorganisms, *Bull. Environ. Contam. Toxicol.*, 13, 666, 1975.
69. **Bollag, J. M. and Liu, S. Y.**, Fungal degradation of 1-naphthol, *Can. J. Microbiol.*, 18, 1113, 1972.
70. **Boush, G. M. and Matsumura, F.**, Insecticidal degradation by *Pseudomonas melophthora*, the bacterial symbiote of the apple maggot, *J. Econ. Entomol.*, 60, 918, 1967.
71. **Matsumura, F. and Boush, G. M.**, Degradation of insecticides by a soil fungus, *Trichoderma viride*, *J. Econ. Entomol.*, 61, 610, 1968.
72. **Williams, I. H., Pepin, H. S., and Brown, M. J.**, Degradation of carbofuran by soil microorganisms, *Bull. Environ. Contam. Toxicol.*, 15, 244, 1976.

73. **Getzin, L. W.**, Persistence and degradation of carbofuran in soil, *Environ. Entomol.*, 2, 461, 1973.
74. **Yu, C. C., Booth, G. M., Hansen, D. J., and Larsen, J. R.**, Fate of carbofuran in a model ecosystem, *J. Agric. Food Chem.*, 22, 431, 1974.
75. **Kandasamy, D., Chendrayan, K., Rajukkannu, K., and Balasubramanian, M.**, *Curr. Sci.*, 46, 280, 1977.
76. **Benezet, H. J. and Matsumura, F.**, Factors influencing the metabolism of mexacarbate by microorganisms, *J. Agric. Food Chem.*, 22, 427, 1974.
77. **MacRae, I. C. and Alexander, M.**, Microbial degradation of selected herbicides in soil, *J. Agric. Food Chem.*, 13, 72, 1965.
78. **Banting, J. D.**, Factors affecting the activity of diallate and triallate, *Weed Res.*, 7, 302, 1967.
79. **Anderson, J. P. E.**, Influence of temperature and moisture on evaporation, degradation, and binding of diallate in soil, *Pflanzenkrankheiten und Pflanzenschutz*, 82, 141, 1975.
80. **Anderson, J. P. E. and Domsch, K. H.**, Microbial degradation of the thio-carbamate herbicide, diallate, in soils and by pure cultures of soil microorganisms, *Arch. Environ. Contam. Toxicol.*, 4, 1, 1976.
81. **Kaufman, D. D.**, Degradation of carbamate herbicides in soil, *J. Agric. Food Chem.*, 15, 582, 1967.
82. **Kearney, P. C. and Kaufman, D. D.**, Enzyme from soil bacterium hydrolyzes phenylcarbamate herbicides, *Science*, 147, 740, 1965.
83. **Kearney, P. C.**, Purification and properties of an enzyme responsible for hydrolyzing phenylcarbamates, *J. Agric. Food Chem.*, 113, 561, 1965.
84. **Geissbuhler, H., Martin, H., and Voss, G.**, The substituted ureas, in *Herbicides — Chemistry, Degradation and Mode of Action*, Vol. 1, 2nd ed., Kearney, P. C. and Kaufman, D. D., Eds., Marcel Dekker, New York, 1973, 209.
85. **Hill, G. D., McGahen, J. W., Baker, H. M., Finnerty, D. W., and Bingeman, C. W.**, The fate of substituted urea herbicides in agricultural soils, *Agron. J.*, 47, 93, 1955.
86. **Ross, J. A. and Tweedy, B. G.**, Degradation of four phenylurea herbicides by mixed populations of microorganisms from two soil types, *Soil Biol. Biochem.*, 5, 739, 1973.
87. **Weinberger, M. and Bollag, J. M.**, Degradation of chlorbromuron and related compounds by the fungus *Rhizoctonia solani*, *Appl. Microbiol.*, 24, 750, 1972.
88. **Wallnofer, P. R. and Bader, J.**, Degradation of urea herbicides by cell-free extracts of *Bacillus sphaericus*, *Appl. Microbiol.*, 19, 714, 1970.
89. **Wallnofer, P.**, The decomposition of urea herbicides by *Bacillus Sphaericus*, isolated from soil, *Weed Res.*, 9, 333, 1969.
90. **Engelhardt, G., Wallnofer, P. R., and Plapp, R.**, Identification of *N,O*-Dimethylhydroxylamine as a microbial degradation product of the herbicide, Linuron, *Appl. Microbiol.*, 23, 664, 1972.
91. **Engelhardt, G., Wallnofer, P. R., and Plapp, R.**, Degradation of Linuron and some other herbicides and fungicides by a Linuron-inducible enzyme obtained from *Bacillus sphaericus*, *Appl. Microbiol.*, 22, 284, 1971.
92. **Engelhardt, G., Wallnofer, P. R., and Plapp, R.**, Purification and properties of an aryl acylamidase of *Bacillus sphaericus*, catalyzing the hydrolysis of various phenylamide herbicides and fungicides, *Appl. Microbiol.*, 26, 709, 1973.
93. **Geissbuhler, H., Haselbach, C., Aebi, H., and Ebner, L.**, The fate of N′-(4-chlorophenoxy)-phenyl-*N,N*-dimethylurea (C-1983) in soils and plants. III. Breakdown in soils and plants, *Weed Res.*, 3, 277, 1963.
94. **Tweedy, B. G., Loeppky, C., and Ross, J. A.**, Metabolism of 3-(*p*-bromophenyl)-1-methoxy-1-methylurea, *J. Agric. Food Chem.*, 18, 851, 1970.
95. **Lanzilotta, R. P. and Pramer, D.**, Herbicide transformation. II. Studies with an acylamidase of *Fusarium solani*, *Appl. Microbiol.*, 19, 307, 1970.
96. **Sharabi, N. El Din and Bordeleau, L. M.**, Biochemical decomposition of the herbicide N-(3,4 dichlorophenyl)-2-methyl pentanimide and related compounds, *Appl. Microbiol.*, 18, 369, 1969.
97. **Bartha, R. and Pramer, D.**, The metabolism of acylanilide herbicides, *Adv. Appl. Microbiol.*, 13, 317, 1970.
98. **Bartha, R.**, Biochemical transformations of anilide herbicides in soil, *J. Agric. Food Chem.*, 16, 602, 1968.
99. **Kaufman, D. D. and Blake, J.**, Microbial degradation of several acetamide, acylanilide, carbamate, toluidine and urea pesticides, *Soil Biol. Biochem.*, 5, 297, 1972.
100. **Beestman, G. B. and Deming, J. M.**, Dissipation of acetanilide herbicides from soils, *Agron. J.*, 66, 308, 1974.
101. **Smith, A. E. and Phillips, D. V.**, Degradation of alachlor by *Rhizoctonia solani*, *Agron. J.*, 67, 347, 1975.
102. **Tiedje, J. M. and Hagedorn, M. L.**, Degradation of alachlor by a soil fungus, *Chaetomium globosum*, *J. Agric. Food Chem.*, 23, 77, 1975.

103. **Chen, Y.-L. and Wu, T.-C.**, Degradation of herbicide butachlor by soil microbes, *J. Pest. Sci.*, 3, 411, 1978.
104. **McGahen, L. L. and Tiedje, J. M.**, Metabolism of two new acylanilide herbicides, Antor herbicide (H-22234) and Dual(metolachlor) by the soil fungus *Chaetomium globosum*, *J. Agric. Food Chem.*, 26, 414, 1978.
105. **Lanzilotta, R. P. and Pramer, D.**, Herbicide transformation. I. Studies with whole cells of *Fusarium solani*, *Appl. Microbiol.*, 19, 301, 1970.
106. **Blake, J. and Kaufman, D. D.**, Characterization of acylanilide-hydrolyzing enzyme(s) from *Fusarium oxysporum* Schecht, *Pest. Biochem. Physiol.*, 5, 305, 1975.
107. **Tiedje, J. M., Duxbury, J. M., Alexander, M., and Dawson, J. E.**, 2,4-D metabolism: pathway of degradation of chlorocatechols by *Arthrabacter* sp., *J. Agric. Food Chem.*, 17, 1021, 1969.
108. **Loos, M. A.**, Phenoxyalkanoic acids, in *Herbicides — Chemistry, Degradation, and Mode of Action*, Vol. 1, 2nd ed., Kearney, P. C. and Kaufman, D. D., Eds., Marcel Dekker, New York, 1975, 1.
109. **Audus, L. J.**, Herbicide behavior in the soil, in *The Physiology and Biochemistry of Herbicides*, Audus, L. J., Ed., Academic Press, New York, 1964, chap. 5.
110. **Kaufman, D. D. and Kearney, P. C.**, Microbial transformations in the soil, in *Herbicides — Physiology, Biochemistry, Ecology*, Vol. 2, Audus, L. J., Ed., Academic Press, New York, 1976, 29.
111. **Hemmett, R. B., Jr. and Faust, S. D.**, Biodegradation kinetics of 2,4-dichlorophenoxyacetic acid by aquatic microorganisms, *Residue Rev.*, 29, 191, 1969.
112. **Loos, M. A., Roberts, R. N., and Alexander, M.**, Phenols as intermediates in the decomposition of phenoxyacetates by an *Arthrobacter* species, *Can. J. Microbiol.*, 13, 679, 1967.
113. **Loos, M. A., Roberts, R. N., and Alexander, M.**, Formation of 2,4-dichlorophenol and 2,4-dichloroanisole from 2,4-dichlorophenoxy-acetate by *Arthrobacter* sp., *Can. J. Microbiol.*, 13, 691, 1967.
114. **Tiedje, J. M. and Alexander, M.**, Enzymatic cleavage of the ether bond of 2,4-dichlorophenoxyacetate, *J. Agric. Food Chem.*, 17, 1080, 1969.
115. **Evans, W. C., Smith, B. S. W., Fernley, H. N., and Davies, J. I.**, Bacterial metabolism of 2,4-dichlorophenoxyacetate, *Biochem. J.*, 122, 543, 1971.
116. **Kilpi, S., Backstrom, V., and Korhola, M.**, Degradation of 2-methyl-4-chlorophenoxyacetic acid (MCPA), 2,4-dichlorophenoxyacetic acid (2,4-D), benzoic acid and salicylic acid by *Pseudomonas* sp. HV3, *FEMS Microbiol. Lett.*, 8, 177, 1980.
117. **Ou, L. T. and Sikka, H. C.**, Extensive degradation of silvex by synergistic action of aquatic microorganisms, *J. Agric. Food Chem.*, 25, 1336, 1977.
118. **Alexander, M. and Aleem, M. I. H.**, Effect of chemical structure on microbial decomposition of aromatic herbicides, *J. Agric. Food Chem.*, 9, 44, 1961.
119. **Horvath, R. S.**, Microbial cometabolism of 2,4,5-trichlorophenoxyacetic acid, *Bull. Environ. Contam. Toxicol.*, 5, 537, 1971.
120. **Rosenberg, A. and Alexander, M.**, Microbial metabolism of 2,4,5-trichlorophenoxyacetic acid in soil, soil suspensions, and axenic culture, *J. Agric. Food Chem.*, 28, 297, 1980.
121. **Rosenberg, A. and Alexander M.**, 2,4,5-trichlorophenoxyacetic acid (2,4,5-T) decomposition in tropical soil and its cometabolism by bacteria in vitro, *J. Agric. Food Chem.*, 28, 705, 1980.
122. **Pemberton, J. M. and Fisher, P. R.**, 2,4-D plasmids and persistence, *Nature (London)*, 268, 732, 1977.
123. **Fisher, P. R., Appleton, J., and Pemberton, J. M.**, Isolation and characterization of the pesticide-degrading plasmid JP1 from *Alcaligenes paradoxus*, *J. Bacteriol.*, 135, 798, 1978.
124. **Loos, M. A., Bollag, J. M., and Alexander, M.**, Phenoxyacetate herbicide detoxication by bacterial enzymes, *J. Agric. Food Chem.* 15, 858, 1967.
125. **Bollag, J. M., Helling, C. S., and Alexander, M.**, 2,4-D metabolism: enzymatic hydroxylation of chlorinated phenols, *J. Agric. Food Chem.*, 16, 826, 1968.
126. **Bollag, J. M., Briggs, G. G., Dawson, J. E., and Alexander, M.**, 2,4-D metabolism: enzymatic degradation of chlorocatechols, *J. Agric. Food Chem.*, 16, 829, 1968.
127. **Bollag, J. M., Helling, C. S., and Alexander, M.**, Metabolism of 4-chloro-2-methyl-phenoxyacetic acid by soil bacteria, *Appl. Microbiol.*, 15, 1393, 1967.
128. **Domsch, K. H., Fischer, H. F., and Munnecke, D. M.**, Mikrobielle bzw. enzymatische spalrung von parathion, in Bundeministerium fur Forschung und Technologie, Final Report, BMFT, Bonn, 1979, 54.
129. **Munnecke, D. M. and Barik, S.**, unpublished data, 1981.
130. **O'Sullivan, D.**, Strong growth ahead for industrial enzymes, *Chem. Eng. News*, 59, 37, 1981.
131. **Goldman, P., Milne, G. W. A., and Keister, D. B.**, Carbon-Halogen bond cleavage. III. Studies on bacterial halidohydrolases, *J. Biol. Chem.*, 243, 428, 1968.
132. **Janke, D. and Fritsche, W.**, Dechlorierung von 4-chlorophenol nach extradioler ringspaltung durch *Pseudomonas putida*, *Z. Allg. Mikrobiol.*, 19, 139, 1979.

Chapter 2

MICROBIAL DEGRADATION OF POLYCHLORINATED BIPHENYLS (PCBs)

Kensuke Furukawa

TABLE OF CONTENTS

I. INTRODUCTION

Polychlorinated biphenyls (PCBs) in wildlife samples were first reported in Sweden in 1966,[1] and later reported in several European countries, North America, and Japan. These widespread and persistent pollutants have become an increased global concern since then, and are recognized to be among the most abundant of the chlorinated hydrocarbons in the global ecosystem as well as 2,2-bis(*p*-chlorophenyl)1,1,1-trichloroethane (DDT) and its metabolites.[2] Research into the ecological and toxicological aspects of PCBs has greatly increased in the last decade; more than 1000 articles have been published so far, according to *Chemical Abstracts.*

PCBs are made by direct chlorination of biphenyl. In the process of replacing hydrogen atoms with chlorines, a large number of substitution combinations arise. For example, 3 monochlorobiphenyl isomers are possible, 12 dichlorobiphenyl isomers, 21 trichlorobiphenyl isomers, and so on. Theoretically, 210 different compounds containing 0 to 10 chlorine atoms per biphenyl molecule can be prepared, but only about half that number are sterically possible. The two rings are numbered identically, except that the ring with the fewest chlorine substituents is designated as primed (Figure 1). PCBs have been manufactured commercially since 1929, and they are marketed under a number of commercial trade names, e.g., Aroclor®, Clophen®, Phenoclor®, and Kanechlor®. The major American producer of PCBs is Monsanto®, and the trademark is Aroclor®. The Aroclor® products are identified by a four digit numbering code in which the first two digits, 12, indicate the parent molecule in biphenyl and the last two digits specify the weight percent of chlorine. The exception is Monsanto's newest product, Aroclor® 1016 (41% chlorine). Aroclor® 1221 is primary monochlorobiphenyl. Similarly, Aroclor® 1232, 1242, 1248, 1254, and 1260 are primary di-, tri-, tetra-, penta-, and hexachlorobiphenyl, respectively. However, Aroclor® 1254, for example, has been reported to contain 69 different chlorinated biphenyl components.[3] Commercial PCBs are thus complex mixtures of chlorinated biphenyls.

The physical and chemical properties that make PCBs desirable industrial chemicals are their excellent thermal stability, their strong resistance to both acidic and basic hydroxides and action of corrosive chemicals, and their general inertness. They are insoluble in water but possess a low finite vapor pressure. Their boiling points range from 278 °C for Aroclor® 1221 to 415°C for Aroclor® 1268. They vary from mobile, oily liquids to white crystals and hard transparent resins.[4] Having such physical and chemical properties, and particularly their flame-retardant characteristics, they are of widespread use in a variety of applications. Major common uses include transformer oils, capacitator dielectrics, heat transfer fluids, hydraulic lubricants, plasticizers, waxes, pesticide extenders, and others.[5] It has been estimated that the total world production of PCBs since 1929 has been close to 2 million tons, of which about half has been produced in the U.S.[6] The PCBs have been steadily released into the environment in many countries, presumably over decades, and are now found to be a pervasive, worldwide contaminant. Industrial sales of PCBs for use in products which would allow leakage to the environment were stopped in the U.S. and Sweden during 1970 to 1971, and from 1972 to 1973 other European countries and Japan initiated similar action. Even with these restrictions, significant quantities of PCBs are still released into the environment.

PCBs have been found in fish, mussels, and birds from the River Rhine and the Netherlands coastal areas; and they have been found in marine animals and wildlife in Sweden, England, Canada, and the U.S.[7-14] A typical PCB residue from fish resembles the Aroclor® 1254 mixture more closely than it does other Aroclors.[15,16] PCBs have also been found in wastewaters,[17,18] household products,[19] drinking water,[20] and in foods,[21,22] particularly in fish[23] and fishery products.[24] It has also been shown that

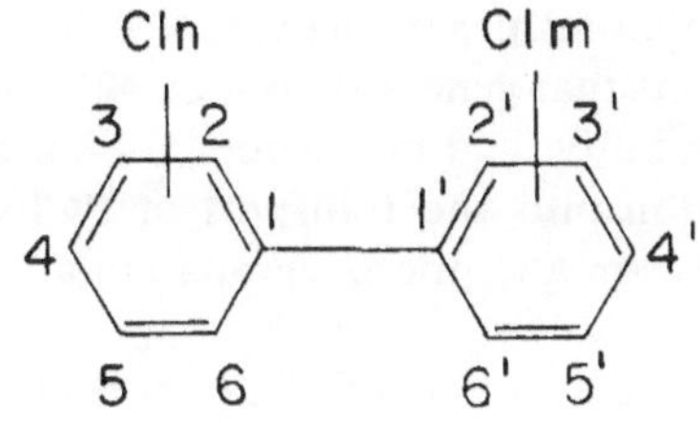

FIGURE 1. The numbering system for the biphenyl structure.

human adipose tissue[25-27] and human milk[27-29] contain PCBs. When PCBs are discharged into river, lake, or sea, they will accumulate on the sediment in relatively high concentration and redissolve very slowly, due to their very low aqueous solubility.[30] The number of PCBs occurring in the environment is of the order of 30 to 40.[31] The main contributors are the various isomers of pentachlorobiphenyl (Aroclor® 1254), hexachlorobiphenyl (Aroclor® 1260), and to a lesser extent, trichlorobiphenyl (Aroclor® 1242). These PCBs tend to concentrate at the trophic level in the biosphere. The determination of PCB residues has been reviewed in detail by Hutzinger et al.[32] and others.[22,33,34]

Acute and subacute toxicity, reproductive effects, mutagenicity, effect on biochemical functions, and metabolism and excretion of both PCB mixtures and individual PCBs on laboratory animals have been extensively studied, and a number of reviews on different aspects of the toxicology of PCBs have been written.[35-39] Nelson et al.[37] discussed extensively the toxic effects produced in fish by PCBs. On the other hand, many studies also have been performed on interaction between PCBs and microorganisms. The toxic effect of PCBs on estuarine and marine phytoplankton, bacteria, and other microbes is now well documented. A number of studies have also indicated that many microorganisms are capable of degrading PCBs.[40-50] Metabolic breakdown by bacteria is now considered to be one of the major routes of environmental degradation for these widespread pollutants.

Factors affecting microbial degradation of PCBs in the environment are

1. The physical and chemical nature of PCB components: degree of chlorination, position of substituted chlorines on the biphenyl molecule, and water solubility and volatility
2. Microbial aspects: occurrence and distribution of PCB degrading organisms, and their ability for PCB degradation
3. Environmental factors: photolysis by sunlight, pH, temperature, oxygen, nutrition, and adsorption of PCBs by such various materials as soil, sediment, sludge, and biological matter

These factors are all combined to varying extents, and affect microbial degradation of PCBs in the ecosystem. The present review describes interaction between PCBs and microorganisms mainly from the viewpoint of biodegradation of PCBs.

II. BEHAVIOR OF PCBs IN THE ECOSYSTEM

A. Transfer of PCBs into the Environment

Since identification of PCB residues in the environment in 1966,[1] an effort has been made to determine the level and distribution of these materials. The PCBs generally found in the environment are the more highly chlorinated biphenyls, i.e., those containing five or more chlorine atoms per biphenyl molecule.[7,8,11] This is in marked contrast to manufactured products that generally contain 42% or less chlorine.[4] The properties of volatility, water solubility, and bioaccumulation are of particular importance as mechanisms of introduction into and transport of PCBs within the environment. The properties of biodegradation and photodegradation are important as mechanisms of removal of PCBs.

Studies on relative volatilities for the PCB components[51] indicate that the lower chlorinated components make up a large part of the total vapor concentration. Significant concentrations (3 to 9 ng/m^3) of Aroclor® 1254 were found in air samples.[52] Laboratory studies by Oloffs et al.[53] showed that Aroclor® 1260 was lost to the atmosphere through volatilization from water. These studies indicate vapor phase transport can be a significant pathway for loss of PCBs from water bodies. Because of the low apparent water solubility, the dissolved concentration of PCBs in aquatic environment would be expected to be low. Saturation of the aqueous solutions of Aroclor® 1242 and 1016 occurred in less than 1 week with the apparent solubility being 0.34 (±0.06) and 0.42 (±0.08) mg/ℓ, respectively.[54] Low chlorinated components are more soluble and contribute more to the water-saturated composition.[51,55] The lipophilic nature of PCBs makes them adsorb microorganisms in aquatic systems.[56-63] The well-known affinity of PCBs for solid surfaces also suggests that adsorption of falling particles could be a significant transfer mechanism.[64]

It has been estimated that 1000 to 2000 tons/year escaped into the atmosphere in past years from plasticized materials containing PCBs and that about 4000 tons/year entered waterways from dumping and leakage of lubricant, hydraulic fluids, and heat transfer fluids.[65] Södergren[66] reported that input of organochlorine residues to the Lake Havgård ecosystem takes place from the following main sources: atmosphere, organisms, drainage and subsoil water from catchment area, and aquatic debris. A recent survey by Greichus et al.[67] for PCB residues in an African lake ecosystem indicated that the levels of PCB concentrations are <0.01 ppm in water, 0.12 ppm in bottom sediment, 0.18 ppm in plankton, 1.0 ppm in oligochaeted, 1.6 ppm in bentric insects, 1.2 to 2.3 ppm in several species of fish, and 1.0 to 13 ppm in some parts of cormorants. The most commonly found PCBs in wildlife samples resemble Aroclor® 1254 and 1260. Veith et al.[16] reported that 13 species of fish taken from 14 regions of Lake Michigan contained 2.7 ppm to 15 ppm PCBs. The most abundant PCBs were tetra-, penta-, hexa-, and heptachlorobiphenyls. Lesser chlorinated biphenyls were present in fish from near-shore waters. More than 80 structurally defined PCBs were identified in marine fish by high resolution glass capillary.[68,69]

Due to the depth of the oceans and their dominance of the surface of the earth, the oceans should be the final repository in the global cycling of PCBs. PCBs were recovered from all surface water and sediment samples collected throughout the Chesapeake Bay and along the southeast Atlantic coast.[70] Elder and Villeneuve[71] reported that PCBs were detectable in all seawater samples collected from 36 locations in the open Mediterranean Sea. The concentrations were within the range of <0.2 to 19 ng/ℓ. Most commonly, concentrations were 1 to 5 ng/ℓ. The environmental accumulation of PCBs within the last 40 years of production has brought a PCB content within the upper 10 to 1000 m of the global hydrosphere. These values of PCB concentrations were measured at places as far apart as the North Atlantic[72] and the open Pacific.[73] Ballshmiter et al.[69] roughly calculated that assuming 3 ng/ℓ as a mean value of PCB concentration in the upper 500 m layer of the global hydrosphere (mean depth 3800 m), the PCB content of this water layer amounts to 0.75×10^6 ton. The overall production since 1929 is estimated to be close to 2×10^6 ton PCBs.[32]

B. Magnification of PCB Concentration in the Environment

The hydrophobic and lipophylic nature of PCBs make them particularly soluble in lipids, and hence PCBs accumulate especially in organisms which are high in the food chain, perhaps including human beings.[25-29] At the microbial level, many organisms are observed to adsorb PCBs. A bacterial strain of *Alcaligenes*[74] was demonstrated to adsorb a large amount of ^{14}C-2,5,2′-trichlorobiphenyl onto the cell surface in a short time. *Aspergillus flavus,* soil fungus, accumulates various Aroclors in the mycelium.[75] Choi et al.[76] observed that a considerable amount of PCBs accumulated in the activated sludges. For example, concentrations in sludge and effluent solution were 36.2 and 0.006 ppm, respectively, after 1 day when Aroclor® 1254 was introduced at the concentration of 16.88 ppm. Urey et al.[57] observed that four PCB isomers and Aroclor® 1242 were concentrated with *Chlorella pyrenoidosa* by several thousandfold whether the cells are alive or dead, suggesting that the mechanism involved is simple partitioning into cell lipids.

Sanders and Chandler[77] demonstrated biological magnification of ^{36}Cl-Aroclor® 1254 from water by aquatic invertebrates such as daphnid, phantom midge, scud, mosquito larvae, grass shrimp, stonefly, dobsonfly, and crayfish. The invertebrates were exposed to Aroclor® 1254 at concentrations less than 3 ppb. For example, scud exposed continuously to 1.6 ± 0.1 ppb of the Aroclor® in water reached an equilibrium concentration in 14 days. Total body residues at this time were 44 ppm, a concentration 27,500 times that in water. Once this equilibrium was reached within the organisms, no further magnification of residue was detected after an additional 21 days exposure. Metcalf et al.[78] surveyed ecological magnification (EM) (parts per million in organisms per parts per million in water) and biodegradability index (BI) (parts per million polar degradation products per parts per million nonpolar products) in several aquatic organisms. The EM value for the parent compounds increased substantially with the number of chlorine atoms, from trichlorobiphenyl to pentachlorobiphenyl. Conversely the BI value decreased with increasing degrees of chlorine. From these observations, it should be emphasized that even a low concentration of PCBs in an aquatic environment can provide a reservoir for bioaccumulation via bacteria, plankton, and higher forms.

C. Toxic Effect of PCBs in the Ecosystem

PCBs discharged in the natural environment have much influence on aquatic and terrestrial organisms, and PCBs are also transformed by them. Many investigations have been conducted on the effects of PCBs on the lower trophic levels of freshwater and marine ecosystems. Keil et al.[79] found inhibition of cell division for a marine diatom *Cylindrotheca closterium* at a concentration of 10 $\mu g/\ell$ (ppb) of Aroclor® 1242 (2 weeks exposure). RNA levels were reduced by 63% and the chlorophyll index reduced by 80% in the organisms. Aroclor® 1254 reduced growth rate and cell population size of two species of marine diatom[80] at a concentration of 10 $\mu g/\ell$ (1 week exposure) and in culture of an estuarine ciliate[81] at a concentration as low as 1 $\mu g/\ell$ (96 hr exposure).

In a heterogenous marine phytoplankton community from the Gulf of Mexico the total radiocarbon uptake was reduced by 50% at 6.5 $\mu g/\ell$ of Aroclor® 1242, or 15 $\mu g/\ell$ of Aroclor® 1254 after a 24 hr exposure.[82] Glooschenko and Glooschenko[81] also found 50% inhibition (2 days exposure) in Great Lake phytoplankton, *Ankistrodesmus faecatus,* at a similar concentration of Aroclor® 1242. Inhibition of carbon fixation was apparently due to reduced level of chlorophyll *a* per PCB-treated cell, but significant loss of function per unit of existing chlorophyll *a* was observed.[83] Such PCB effects upon cell division appear to be species dependent. Mosser et al.[84] studied the effect of PCBs upon five algal species including two diatoms, a marine green alga, and two freshwater species. The marine diatoms showed decreased growth rates between 10 and 100 μg/ℓ, while the other species tolerated higher concentrations. Aroclor® 1254 in concentrations up to 100 μg/ℓ added to mixed cultures of a marine diatom and a marine alga resulted in the loss of dominance by the diatom, suggesting that natural species relationships can be disrupted by these chlorinated hydrocarbons. A low level of PCBs (0.1 μg/ℓ) is more toxic to mixed cultures of algae than to single species in culture.[85]

With Aroclor® 1221, 1232, and 1242, the toxicity to *Euglena gracilis* (freshwater phytoplankton) decreased with increasing chlorination of the Aroclor®.[86] Hawes et al.[87] also observed that toxicity of Aroclors to *Chlorella pyrenoidas* (green algae) decreased in the order of Aroclor® 1242, 1254, and 1268. The similar effect was observed on the motility of the green algae *Phormidium cyanophyceae.*[88] Mono-, di-, and trichlorobiphenyls inhibited the motility of the algal species, while higher chlorinated biphenyls had no effect. Craigie and Hutzinger[89] showed that the biological response depended upon the molecular structure (in the experiment with biphenyls containing identical percentages of chlorine). The 2,4,2′,4′-isomer was more toxic to certain species of phytoplankton than the 2,5,2′,5′-isomer, and both were more deleterious than either the 2,3,4,5- or 3,4,3′,4′-isomers. The latter was not toxic even when added at 100 μg/mℓ (ppm).

In general, most marine diatoms are very sensitive to PCBs, whereas algae show variable degrees of resistance. The differential sensitivity of phytoplanktons might result in alteration in species composition of communities, and this could result in deleterious effects on estuarine and marine ecosystems if the persistent species that become dominant were an inferior food source for the resident herbivores. It has also been postulated that low chlorinated PCBs have much influence on growth kinetics of phytoplankton cultures and that the biological response depends upon the molecular structures of PCB isomers. Thus much is known about the effect of PCBs on phytoplanktons, but there is little information available regarding the effects that phytoplanktons have upon PCBs, i.e., their biotransformation.

The biological interaction between a series of PCBs and dieldrin and DDT was studied by Lichtenstein et al.[90] Many PCBs were found toxic to some insects, but to a lesser extent than dieldrin or DDT. Their toxicity increased with a decrease in their chlorine content and, moreover, nontoxic amounts of PCBs increased the toxicity of dieldrin and DDT. Potential effects of PCBs on biological systems, especially in combination with other synthetic chemicals, should thus be considered. Yoshimura et al.[91] showed that acute toxicity of 5-hydroxy-2,4,3′,4′-tetrachlorobiphenyl, a metabolite, was five times higher than the parent 2,4,3′,4′-tetrachlorobiphenyl in male mice.

Lower chlorinated biphenyls are of special interest because their biological degradability might lead to metabolites of increased toxicity. Recent work has shown that the bihydroxylation of PCBs by mammals proceeds via arene oxide intermediates,[92,93] and these activated aromatic molecules are associated with chemical carcinogenicity.[94] Furukawa et al.[95] observed that various known and unknown metabolites such as dihydroxy compounds, ring-cleavage compounds, and chlorobenzoic acids with varying chlorine contents were produced from commercial PCB mixtures by bacteria. Furthermore, there is evidence that some commercial PCBs, those manufactured in the U.S., European countries, and Japan[96-98] may contain trace amounts of chlorinated dibenzofurans or other extensively toxic impurities. Crosby and Moilanen[99] reported that irradiation of chlorobiphenyls dispersed in water revealed an identification trace of 2-chlorodibenzofuran as a photolysis product of 2,5-dichlorobiphenyl and 2,5,2′,5′-tetrachlorobiphenyl in roughly a 0.2% steady-state yield. Morita et al.[100] also reported the formation of polychlorinated dibenzofurans from PCB mixtures by heating at

Table 1
RELEASE OF METABOLITES FROM CELL-BOUND ^{14}C-2,5,2′-TRICHLOROBIPHENYL IN *ALCALIGENES* sp. Y42

Incubated for (hr)	Supernatant cpm (%)[a]	Cell cpm (%)[a]	Total cpm (%)	Recovery (%)[b]
0.5	6282 (5.1)	116184(94.9)	123146(100)	71
2	7524 (6.0)	117132(94.0)	124656(100)	72
4	14142(10.3)	123712(89.7)	137856(100)	79
6	18246(12.3)	130283(87.7)	148529(100)	85
8	19782(14.0)	121072(86.0)	140854(100)	81
24	31356(20.1)	124470(79.9)	155826(100)	90

Note: 100 μg of ^{14}C-2,5,2′-trichlorobiphenyl (0.1 μCi) was incubated with *Alcaligenes* sp. Y42 (2.0×10^9 cells per mℓ) in 10 mℓ 0.05 *M* phosphate buffer (pH 7.5) for predetermined periods, and radioactivities of the supernatant and the cells were measured.

[a] Percentage for total cpm recovered.
[b] Percentage recovery for original used 0.1 μCi.

From Furukawa, K., Matsumura, F., and Tonomura, K., *Agric. Biol. Chem.*, 42, 542, 1978. With permission.

300°C for 1 week with 0.2% yield. Knowledge of the influence on living organisms of biological metabolites or nonbiological products from PCBs should be extensively accumulated to better interpret their environmental hazard.

III. INTERACTION BETWEEN PCBs AND MICROORGANISMS

Interaction between PCBs and bacteria has mainly been investigated from the viewpoint of bacterial degradation of PCBs. A biphenyl-assimilating bacterium was first reported by Lunt and Evans.[101] The Gram-negative bacterium oxidized biphenyl to phenylpyruvic acid. *Pseudomonas putida* metabolized biphenyl to benzoic acid.[102,103] Ohmori et al.[104] isolated 39 strains of microorganisms capable of assimilating biphenyl by repeating liquid enrichment techniques. In 1973 Ahmed and Focht[40] first reportedly showed that two species of *Achromobacter* were capable of growing on biphenyl and 4-chlorobiphenyl degraded mono- and dichlorobiphenyl to chlorobenzoic acids. Chlorobenzoic acids were shown to be metabolites of both pure chlorobiphenyls and commercial mixtures, when these compounds were degraded by certain pure strains or by mixed cultures of bacteria. Furukawa et al.[46,74] demonstrated the interaction between PCBs and bacteria. An *Alcaligenes* sp. and *Acinetobacter* sp. isolated from lake sediment were able to adsorb a large amount of ^{14}C-2,5,2′-trichlorobiphenyl onto the cell surface in a short time and then gradually metabolized this. The metabolites such as ring-cleavage product and 2,5-dichlorobenzoic acid were moved into the aqueous layer from the cells with time (Table 1 and Figure 2.) In the mixed culture, interaction between the two organisms was observed. Both *Alcaligenes* sp. and *Acinetobacter* sp. were able to grow on biphenyl and 4-chlorobiphenyl in pure culture, but when the two organisms were incubated with biphenyl as a sole carbon source, *Alcaligenes* sp. became dominant due to its rapid growth. On the other hand, 4-chlorobiphenyl was used as a sole carbon source, *Acinetobacter* sp. was always dominant at the final stage of incubation, and a large amount of 4-chlorobenzoic acid accumulated in the medium. Keil et al.[105] reported that 0.01 and 0.1 ppm Aroclor® 1242 consistently stimulated growth of *Escherichia coli* in vitro. Sayler et al.[48] found that a *Pseudomonas* sp. de-

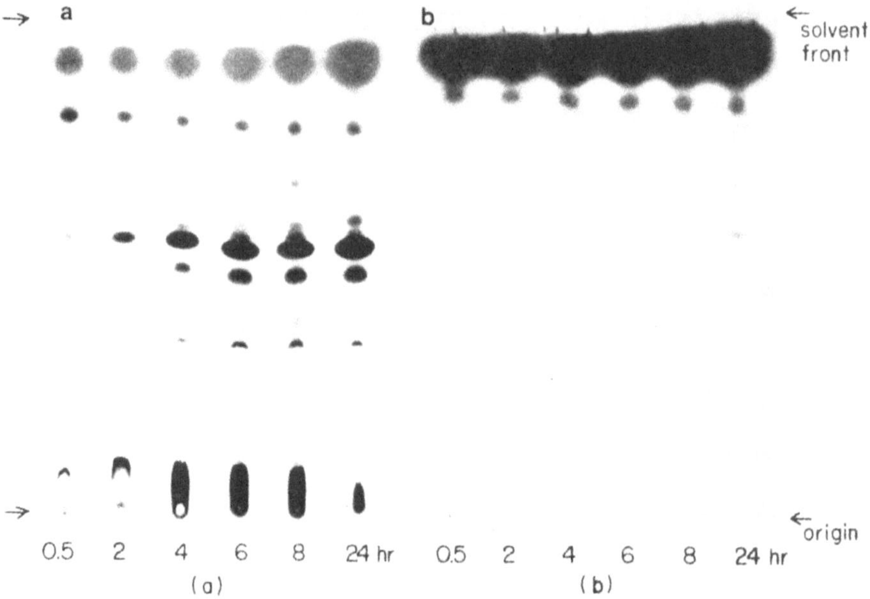

FIGURE 2. Representation of the autoradiograms of (a) the supernatant extracts and (b) the cell 80% ethanol extracts as separated on a silica gel thin layer chromatography (TLC). The TLC plates were developed by using benzene-dioxane-acetic acid (90:20:4). (From Furukawa, K., Matsumura, F., and Tonomura, K., *Agric. Biol. Chem.*, 42, 542, 1978. With permission.)

graded Aroclor® 1254 and hexachlorobiphenyl. In addition to apparent degradation of PCBs, it was observed that PCBs stimulated growth and oxygen uptake.

PCB-degrading bacteria seem to be distributed commonly in the environment. The bacteria recovered from estuarine and marine samples represented seven bacterial genera: *Pseudomonas, Vibrio, Aeromonas, Micrococcus, Acinetobacter, Bacillus*, and *Streptomyces.*[70] Twenty isolates of PCB-degrading bacteria from the Hudson River represented five genera: *Acetobacter, Acinetobacter, Alcaligenes, Klebsiella,* and *Pseudomonas.*[106] Thus a number of bacteria are able to degrade PCBs or are resistant to PCBs; however PCB-sensitive bacteria also exist. Bourquin and Cassidy[107] examined the effect of PCBs on the growth of estuarine bacteria. Of 85 different isolates, growth of 26 strains was inhibited to varying extents by PCBs. In contrast to insensitive isolates, sensitive strains were mainly amylolytic and proteolytic. Blakemore and Carey[108] found that of 17 bacterial cultures tested, growth of only 2 open ocean isolates including one *Pseudomonas* was consistently inhibited by Aroclor® 1254 at concentrations as low as 10 $\mu g/\ell$. The *Pseudomonas* also had reduced carotenoid levels and an altered filamentous morphology. The effect of Aroclor® 1254 was immediate and reversible for a period of at least 18 hr after initial exposure. The reversibility of the effects of PCBs on growth rates and cell yields indicates that these compounds are bacteriostatic for the cells. The PCBs had no effect on respiration rates and protein synthesis at the growth-inhibiting concentration of the Aroclor®. On the other hand, rates per cell of adenine uptake and adenine incorporation into DNA and total nucleic acids by the cells treated with PCBs were significantly lower than in control cells. The effect of PCBs on nucleic acid synthesis may involve a complex regulatory process associated with transport and accumulation of adenine and perhaps other required precursors. Less sensitive bacteria may be more efficient at compensating for impaired uptake of these preformed exogenous nutrients, e.g., via *de novo* synthesis. The apparent transport and oxidation of selected substrates suggest the possibility that PCBs might interfere with the biosynthetic rather than catabolic capacity of sensitive bacteria. The hydrophobic nature of the chlorinated hydrocarbons presumably enhances their toxic properties by promoting efficient binding to cells. Wyndham et al.[109] showed that the low-chlorinated PCBs were highly mutagenic to *Salmonella typhimurium* whereas higher chlorinated ones were only weakly mutagenic.

There is relatively little information about interaction between PCBs and fungi or yeasts, compared with bacteria. Murado et al.[75] studied interaction between PCBs and soil fungus, *Aspergillus flavus.* They observed the following effects: (1) no metabolic change of PCBs was observed, (2) accumulation of various Aroclors was observed in the mycelium, (3) growth was suppressed in the presence of PCBs, (4) nucleic acid content was not significantly affected, and (5) in cultures on solid media there appeared an obvious sporulation delay and a decrease of diameter in colonies. The magnitude of these effects was highest for Aroclor® 1232 and lowest for Aroclor® 1260. There is some evidence that fungi metabolize biphenyl to metabolites similar to those formed by mammalian systems. Smith and Rosazza[110] and Dodge et al.[111] showed that species of several genera of fungi formed mono- and dihydroxybiphenyls. Mono- and dichlorobiphenyl were also transformed to hydroxylated compounds by soil fungus, *Rhizopus japonicus.*[41] Tejedor et al.[112] examined the effect of PCBs on *Saccharomyces cerevisiae* and they found that growth rate and oxygen uptake were suppressed by PCBs, and these effects increased as the chlorine content of PCBs decreased. Wiseman et al.[113] presented evidence that microsomal preparations from *Candida tropicalis* oxidized biphenyl to 4-hydroxybiphenyl.

IV. EFFECT OF CHLORINE SUBSTITUTION ON BIODEGRADABILITY OF PCBs

There is evidence that biological attack is slowed down by insertion of chlorine on the aromatic moiety. Because chlorine is such a strong electronegative substituent, it is known to have a profound effect on chemical reactivity. Chlorine is known to alter the resonant properties of aromatic substrates and in turn alter the electron density of specific sites. These alterations may result in a deactivation of the primary oxidation of chlorobiphenyls by microorganisms. This phenomenon is also observed in the oxidative mechanism of chlorobenzoic acids.[114] Furthermore, the position of substituted chlorines will have stereochemical effects on the affinity between enzymes and substrate molecules. PCBs were at first considered to be more stable than DDT and its metabolites since they lack the ethane component between the aromatic rings, which is the site of action in DDT degradation.[115] However, environmental monitorings have demonstrated that the PCBs generally found in the weathered biological and environmental samples are the more highly chlorinated biphenyls, i.e., those containing five or more chlorine atoms per biphenyl molecule.[7,8,11] This is strong evidence that the less chlorinated materials degrade more rapidly than the more highly chlorinated ones.

Since Ahmed and Focht[40] first showed PCB degrading *Achromobacter,* many workers confirmed that both pure chlorobiphenyls and commercial PCBs are degraded by certain pure strains or by mixed cultures. Degradation of highly chlorinated biphenyls by bacteria can be considered to be a co-metabolic mechanism.[116] Baxter et al.[44] observed that the degradation of PCBs by a *Nocardia* sp. and a *Pseudomonas* sp. increased by the addition of biphenyl. Mixed cultures including predominantly *Alcaligenes odrans, A. denitrificans,* and an unidentified bacterium degraded Aroclor® 1242.[50] The gas chromatography (GC) profile of Aroclor® 1242 was greatly altered after 5 days incubation as shown in Figure 3. Tucker et al.[43] demonstrated that commercial PCB mixtures which contain predominantly mono- and dichlorobiphenyls

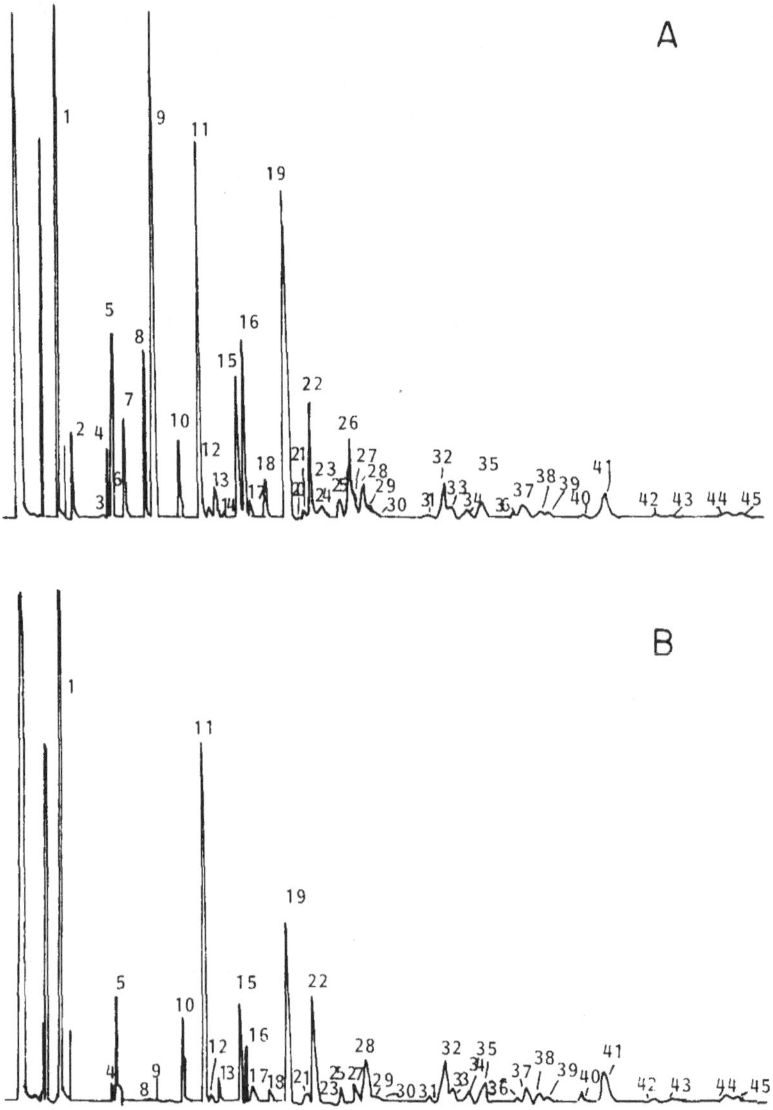

FIGURE 3. Comparison of capillary column gas chromatograms from (A) 5 day blank and (B) 5 day degradation samples using mixed microbial culture. (From Clark, R. R., Chian, E. S. K., and Griffin, R. A., *Appl. Environ. Microbiol.*, 37, 680, 1979. With permission.)

readily undergo primary biodegradation by activated sludge microorganisms, and that as the levels of tri-, tetra-, and pentachlorobiphenyls increase, the degradation rates decrease accordingly (Figure 4). Similar results were obtained for the degradation of Kaneclor® (KC) 200, 300, 400, and 500, which are primary di-, tri-, tetra-, and pentachlorobiphenyl, respectively.[95] Many of the isomeric PCB peaks of KC® 200 in the profile on GC were rapidly decreased after 2 hr incubation with the organisms. The GC profiles of KC® 300 and KC® 400 were greatly altered by the decrease of PCB components and concomitant formation of various metabolites, whereas KC® 500 was very refractile, and the GC profile was not significantly altered even after 4 days incubation.

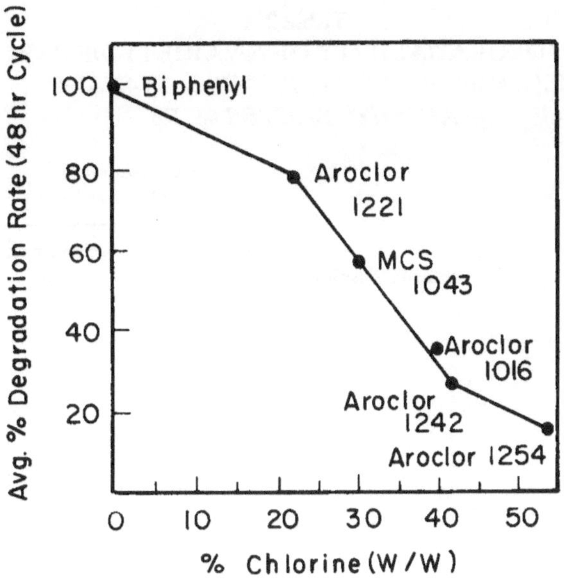

FIGURE 4. Primary biodegradation rates of commercial PCBs as a function of the weight percent chlorine. Addition rate, 1 mg/48 hr. (From Tucker, E. S., Saeger, V. W., and Hicks, O., *Bull. Environ. Contam. Toxicol.*, 14, 705, 1975. With permission.)

Furukawa et al.[117] studied the relation of chemical structures to microbial breakdown of PCBs. Chlorinated biphenyls can theoretically present as many as 210 different components. Of possible isomers, 103 are considered most probable.[118] For this study 31 pure isomers of PCBs and two bacterial strains of *Alcaligenes* sp. and *Acinetobacter* sp. were employed. The former was grown on biphenyl, and the latter on 4-chlorobiphenyl as a sole source of carbon. These organisms could co-metabolize a variety of PCB components to varying extents. The results are presented in Table 2. From the results obtained in this experiment, the relationship between structures of PCBs and their biodegradability is illustrated as follows:

1. Degradation decreased as chlorine substitution increased. The resistance of the more highly chlorinated atoms per molecule, especially more than four chlorines, may explain their detection as residues in the environmental samples.[7,8,11]
2. PCBs containing two chlorines in the ortho position of a single ring (i.e., 2,6) and in each ring (i.e., 2,2′-) showed striking resistance to degradation, as illustrated in the studies with 2,6-, 2,2-, 2,3,6-, 2,5,2′-, 2,3,5,6-, and 2,6,2′,6′-chlorobiphenyls. 2,4,6-Trichlorobiphenyl was the exception in this series of PCBs. This compound was rapidly metabolized by *Acinetobacter* sp., while *Alcaligenes* sp. slowly metabolized the same compound (Figure 5).
3. PCBs containing all chlorine atoms on a single ring were generally degraded faster than those containing the same number on both rings.
4. Further studies[119] have shown that PCB isomers having two chlorine atoms at the 2,3-position of one ring such as 2,3,2′,3′-, 2,3,2′,5′-, and 2,4,5,2′,3′-chlorobiphenyls were much more susceptible to microbial degradation compared with other tetra- and pentachlorobiphenyls.
5. Preferential ring fission of the molecules occurred with a nonchlorinated or less chlorinated ring.

Table 2
BIODEGRADABILITY OF A VARIETY OF PCB ISOMERS BY *ALCALIGENES* sp. Y42 AND *ACINETOBACTER* sp. P6

		Degradation rate (nmol/m*l* per hr)[a]	
Isomer group	PCB chlorine position[b]	Alcaligenes Y42[c]	Acinetobacter P6[c]
A	2-	>50	>50
	3-	>50	>50
	4-	>50	>50
B	2,3-	>50	46.4
	2,4-	>50	>50
	2,5-	>50	>50
	2,6-	0	4.1
	3,4-	>50	44.1
	3,5-	>50	48.2
C	2,2′-	6.3	14.0
	2,4′-	48.2	49.1
	3,3′-	—	18.5
	4,4′-	16.2	25.2
D	2,3,4-	35.1	32.0
	2,3,6-	0	0
	2,4,5-	46.0	32.4
	2,4,6-	3.1	46.0
E	2,5,2′-	1.6	5.1
	2,5,3′-	42.1	41.3
	2,5,4′-	21.8	30.4
	2,4,4′-	41.3	40.2
	3,4,2′-	15.6	38.6
F	2,3,4,5-	25.8	19.0
	2,3,5,6-	0	0
G	2,3,2′3′-	8.7	7.3
	2,4,2′4′-	0	0
	2,4,3′4′-	0	0
	2,5,2′,5′-	0	3.5
	2,6,2′,6′-	1.4	0
	3,4,3′,4′-	0	0
H	2,4,5,2′,5′-	0.6	0

[a] Degradation rate was calculated after 1 hr of incubation for mono-, di-, and trichlorobiphenyls and after 2 hr of incubation for tetra- and pentachlorobiphenyls.

[b] A 500-nmol amount of each compound dissolved in ethanol (10 m*M* each) was suspended in 10 m*l* of phosphate buffer (pH 7.5).

[c] The cell number was 2×10^9/m*l* in *Alcaligenes* Y42 and 4.4×10^8/m*l* in *Acinetobacter* P6.

From Furukawa, K., Tonomura, K., and Kamibayashi, A., *Appl. Environ. Microbiol.*, 35, 223, 1978. With permission.

It is also known that chlorine substitution greatly affects metabolism and excretion of PCBs by mammalian and avian species. A number of less chlorinated PCBs (i.e., mono-, di-, tri-, and tetrachlorobiphenyls) are readily metabolized and excreted by mammals and birds.[120-123] However, the rate of metabolism drops sharply as the number of chlorines per molecule increases.[124,125] For example, the rate of metabolism and

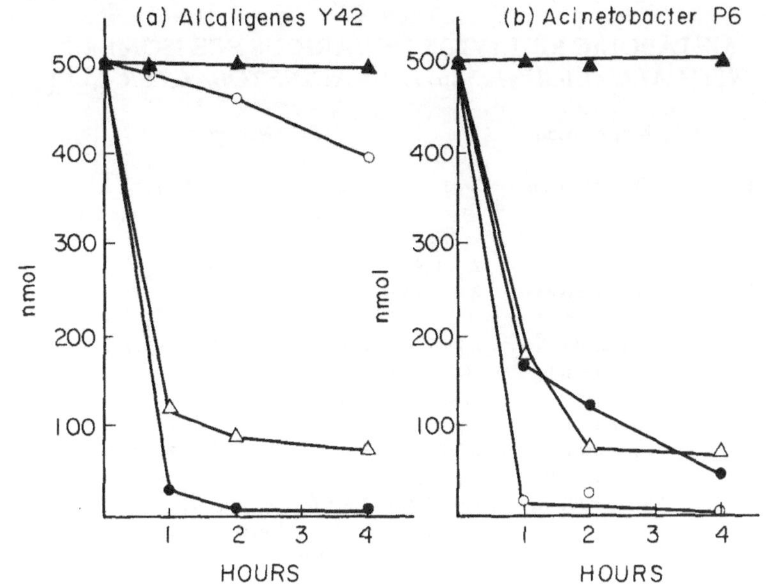

FIGURE 5. Degradation of trichlorobiphenyls by (a) *Alcaligenes* sp. Y42 and (b) *Acinetobacter* sp. P6. Symbols: 2,3,4- (△), 2,3,6- (▲), 2,4,5- (●), and 2,4,6- (○) trichlorobiphenyl. (From Furukawa, K., Tonomura, K., and Kamibayashi, A., *Appl. Environ. Microbiol.*, 35, 223, 1978. With permission.)

excretion of several PCB isomers has been shown to decrease in the following order: 4-chlorobiphenyl, 4,4′-dichlorobiphenyl, 2,5,2′,5′-tetrachlorobiphenyl, 2,4,5,2′,5′-pentachlorobiphenyl.[124] The rate of metabolism and excretion of 4,4′-dichlorobiphenyl is approximately one half that of 4-chlorobiphenyl. Hydroxylation of chlorinated biphenyls in mammals appears to occur through an arene oxide intermediate[92,93] requiring two adjacent unsubstituted carbon atoms.[126] Therefore, the presence of two unsubstituted carbon atoms on the biphenyl ring markedly enhances the metabolic rate constant. For example, the initial half-life of 2,4,5,2′,4′,5′-hexachlorobiphenyl appears to be infinite, whereas the initial half-life of 2,4,5,2′,5′-pentachlorobiphenyl is only 2 days.[124] The position of chlorine substitution also affects the retention and elimination of single homologs.[127]

V. METABOLISM OF PCBs BY MICROORGANISMS

In bacterial metabolism studies with 36 pure isomers (chlorine number 1 to 5) of PCBs, Furukawa et al.[119,128] demonstrated that 23 compounds were metabolized by an *Alcaligenes* sp., and 33 compounds were metabolized by an *Acinetobacter* sp. The 36 PCB isomers examined were divided into six groups according to the pattern of metabolic behavior as presented in Table 3. The following is a further explanation of findings for each of these groups.

1. From the analysis of metabolic intermediates, it was found that a number of PCB components listed in Table 3 were converted to corresponding chlorobenzoic acids by an oxidative route as illustrated in Figure 6. Like other aromatic hydrocarbons, PCBs are considered to proceed via a dioxygenase-catalyzed reaction.[129-131] An initial oxidation process of PCBs has not yet been investigated, but Gibson et al.[132] previously demonstrated a species of *Beijerinckia* metabolized

Table 3
METABOLIC BEHAVIOR OF VARIOUS PCB ISOMERS WITH *ALCALIGENES* sp. AND *ACINETOBACTER* sp.[119]

Group	Major product	PCB isomers
1	Chlorobenzoic acids	2-, 3-, 4-, 2,3-, 2,4-, 2,5-, 3,4-, 3,5-, 2,2′-, 3,3′-, 4,4′-, 2,3,4-, 2,4,5-, 2,5,2′-, 3,4,2′-, 2,4,6-(*Alcaligenes* sp.) 2,4,2′,4′-, 3,4,3′,4′-(*Acinetobacter* sp.)
2	Meta-cleavage products	2,4′-, 2,4,4′-, 2,5,4′-
3	Dihydroxy compounds	2,6-, 2,3,6-, 2,4,2′,5′-, 2,5,2′,5′-, 2,4,5,2′,5′-(*Acinetobacter* sp.)
4	Trihydroxy compounds	2,4,6- (*Acinetobacter* sp.)
5	Unknown compounds	2,3,2′,3′-, 2,3,2′,5′-, 2,4,5,2′,3′- (*Acinetobacter* sp.)
6	No metabolite	2,6-, 2,3,6-, 2,3,5,6-, 2,4,2′,4′-, 2,4,2′,5′-, 2,5,2′,5′-, 2,6,2′,6′-, 2,4,3′,4′-, 2,5,3′,4′-, 3,4,3′,4′-, 2,3,4,5,6-, 2,4,5,2′,3′-, 2,4,5,2′,5′-(*Alcaligenes* sp.) 2,3,5,6-, 2,6,2′,6′-, 2,3,4,5,6- (*Acinetobacter* sp.)

biphenyl to *cis*-2,3-dihydroxy-1-phenylhexa 4,6-diene. This suggests that a cyclic peroxide is an initial oxidation product in the metabolism of biphenyl. Thus two atoms of molecular oxygen would most likely be incorporated at the 2,3-position of the lesser chlorinated ring of PCBs to form *cis*-dihydrodiol compounds (II in Figure 6). Then the dihydrodiols will be dehydrogenated to yield 2′,3′-dihydroxy compound (III). The meta-cleavage will occur at the position between C1 and C2 of the dihydroxy compound. The yellow meta-cleavage products were usually converted rapidly to chlorobenzoic acids in many PCB components. Neither the ortho-cleavage between C2 and C3 nor the meta-cleavage between C3 and C4 of the dihydroxy compounds was found in the experiment. Of 36 PCBs examined, 18 isomers were converted to chlorobenzoic acids through this major metabolic process by *Acinetobacter* sp. and/or *Alcaligenes* sp. Chlorinated benzoic acids were found in the bacterial metabolism of both isomeric chlorobiphenyls and commercial PCB mixtures by several workers.[40,47,104] Therefore, this oxidative route is considered to be common for bacterial degradation of PCBs which have two unsubstituted carbon atoms at the 2,3-position of one of the rings.

2. 2,4′-Di-, 2,4,4′- and 2,5,4′-trichlorobiphenyls which all have chlorine substituents at the 2,4′-position are converted to the meta-cleavage compounds through the major pathway illustrated in Figure 6. These meta-cleavage products are stable and accumulated in the reaction mixtures for further incubation. A dechlorination product of dichlorodihydroxy biphenyl was also observed in the metabolism of 2,4,4′-trichlorobiphenyl. Various intermediates found in 2,4,4′-trichlorobiphenyl metabolism by *Acinetobacter* sp. are presented in Figure 7.[128]
3. Dihydroxy compounds accumulated in the reaction mixture from the 2,6-, 2,3,6-, 2,4,2′,5′-, 2,5,2′,5′-, and 2,4,5,2′,5′-chlorobiphenyls by *Acinetobacter* sp. The position of hydroxyl has not yet been clear, but the 2′,3′-position is not likely for 2,5,2′,5′-tetra- and 2,4,5,2′,5′-pentachlorobiphenyls since all the ortho positions of both rings are occupied by chlorine atoms. The 3′,4′-position is most likely.
4. 2,4,6-Trichlorobiphenyl was quite differently metabolized between *Alcaligenes*

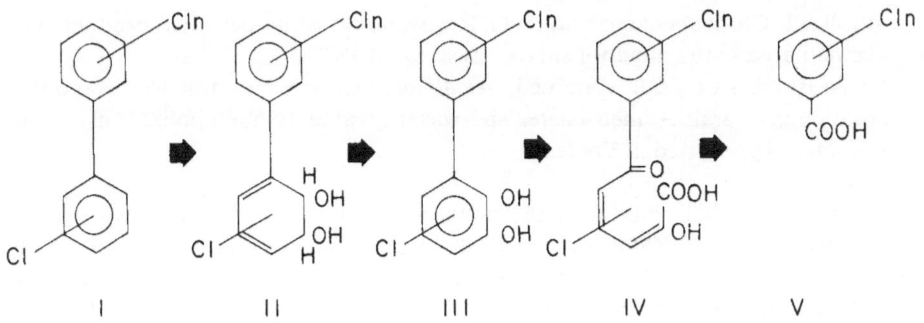

FIGURE 6. Major metabolic sequence of PCBs by *Alcaligenes* sp. Y42 and *Acinetobacter* sp. P6. n = 1 to 4. (From Furukawa, K., Tomizuka, N., and Kamibayashi, A., *Appl. Environ. Microbiol.*, 38, 301, 1979. With permission.)

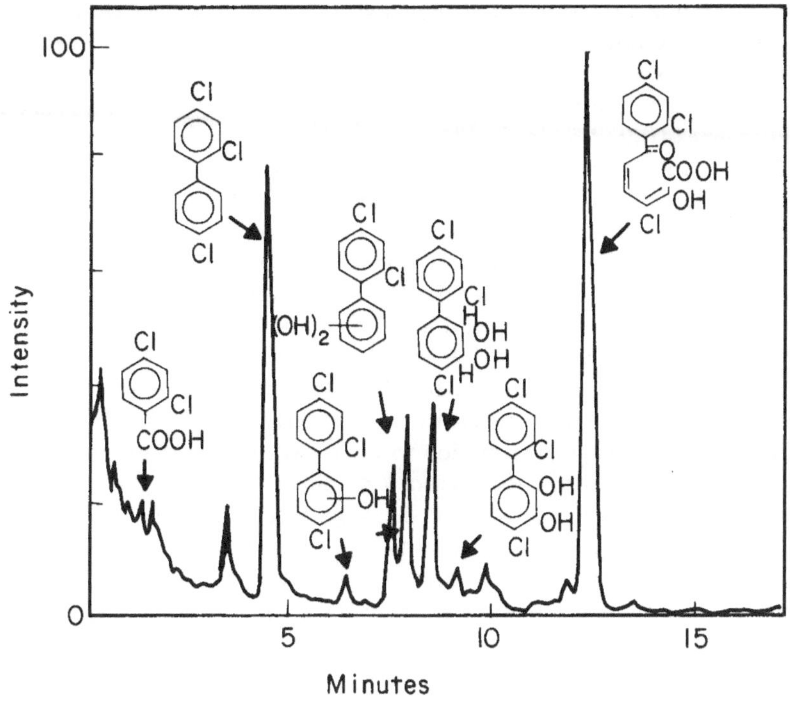

FIGURE 7. Total ion chromatogram on GC-EI-MS of 2,4,4′-trichlorobiphenyl metabolites by *Acinetobacter* sp. P6. (From Furukawa, K., Tonomura, K., and Kamibayashi, A., *Agric. Biol. Chem.*, 43, 1577, 1979. With permission.)

sp. and *Acinetobacter* sp. *Alcaligenes* sp. metabolized this compound very slowly to trichlorobenzoic acid through the major pathway. On the other hand, *Acinetobacter* sp. rapidly converted the same compound to a trihydroxy compound via a dihydroxy compound. Trihydroxy compounds were observed as a minor metabolite in 2,3-, 2,6-, 2,3,6-, and 2,3,4,5-chlorobiphenyls by *Acinetobacter* sp. However, the one from 2,4,6-trichlorobiphenyl was predominant in a late stage of incubation.

5. PCBs which possess chlorine atoms at the 2,3-position in the molecule, such as 2,3,2′,3′-, 2,3,2′,5′-, and 2,4,5,2′,3′-chlorobiphenyl seem to be alternatively me-

tabolized. Chlorobenzoic acid and a large amount of unidentified product always accumulated in the metabolism of this series of PCBs.

6. Of 36 isomers of PCB examined, *Alcaligenes* sp. was not able to metabolize 13 compounds, and *Acinetobacter* sp. was not able to metabolize only 3 compounds, as presented in Table 3.

For commercial PCB mixtures, the metabolic products were examined using the same bacterial strains.[95] Kaneclor® (KC) 200 was rapidly degraded and monochlorobenzoic acids predominantly accumulated in the reaction mixture after 2 hr incubation. Many components of KC® 300 and KC® 400 were metabolized by *Acinetobacter* sp., and mono-, di-, and trichlorobenzoic acids, trichloro- and tetrachlorodihydroxy compounds, the meta-cleavage products containing two to four chlorines were detected by GC-MS analysis. A large amount of the unknown products which may be derived from 2,3,2′,3′- and/or 2,3,2′,5′-tetrachlorobiphenyls were also detected in the metabolism of KC® 400. KC® 500 was refractile and hardly metabolized by the two organisms. Dihydroxy compounds of pentachlorobiphenyl, however, was detected after 1 day incubation. Thus metabolic behavior of PCBs by bacteria are greatly dependent on chlorine substitution.

Kaiser and Wong[133] reported the formation of completely dechlorinated alkanes and alkylated benzenes from Aroclor® 1242 by aquatic organisms; however, this has not been confirmed. Carey and Harvey[49] reported that mixed culture of marine bacteria are capable of metabolizing PCBs to an acid lactone whose structure was partially characterized. Several studies have been conducted on metabolism of biphenyl by bacteria. Lunt and Evans[101] reported that a Gram-negative bacterium assimilated biphenyl and produced 2,3-dihydroxy biphenyl. The compound was further converted to phenylpyruvate via α-hydroxy β-phenyl muconic semialdehyde (a meta-cleavage product between the C3 and C4 positions of 2,3-dihydroxy biphenyl). Ohmori et al.[104] found that certain bacterial strains produced a large amount of γ-benzoyl butyric acid from biphenyl. Catelani et al.[103] studied the metabolism of biphenyl by *Pseudomonas putida*. Biphenyl underwent oxygenation to 2,3-dihydro-2,3-dihydroxy biphenyl. The dihydrodiol was converted to 2,3-dihydroxy biphenyl, then ring fission occurred between the C1 and C2 positions. The meta-cleavage product was then converted to benzoic acid. Consequently, the major pathway of PCBs by bacteria illustrated in Figure 6 seems to be similar to that of metabolism of biphenyl by *P. putida*.

There is some information about metabolism of biphenyl and low chlorinated biphenyls by fungi and yeasts. Fungal species of several genera were able to grow on biphenyl and formed 2- and 4-hydroxybiphenyls and 4,4′-dihydroxybiphenyl.[110] Dodge et al.[111] demonstrated that *Cunninghamella elegans* grown on dextrose broth transformed biphenyl to produce 2-, 3-, and 4-hydroxybiphenyls as well as 4,4′-, 2,4′-, and 2,5-dihydroxybiphenyls. Wallnöffer et al.[41] observed that *Rhizopus japonicus* converted 4-chlorobiphenyl to 4-chloro-4-hydroxybiphenyl. There is only a report that *Candida tropicallis* oxidizes biphenyl to 4-hydroxybiphenyl.[113]

In the natural environment a mixed culture acting on multicomponents of PCB might be expected to give more complete degradation than do single organisms. Chlorinated benzoic acids formed from many PCB components by bacteria can be decomposed by other bacteria. Many workers have studied the degradation of chlorobenzoic acids by both pure cultures of microorganisms and naturally occurring microbial populations. Horvath and Alexander[134] demonstrated that 20 isolates representing 9 bacterial genera were shown to co-metabolize one or more of 22 substituted benzoates. Some of the chlorinated benzoates can be acted by pure cultures of such genera as *Arthrobacter*,[135] *Brevibacterium*,[136] *Alcaligenes*,[114] and *Pseudomonas*,[137,138] and the metabolic processes are studied in detail. Transformation and fate of chlorobenzoates under natural conditions has also been studied by some workers. Dewey et al.[139] observed biological degradation of 2,3,6-trichlorobenzoic acid in soil. GiGeronimo et al.[140] reported that microorganisms in sewage decomposed *o*-, *m*-, *p*-chlorobenzoates and 3,4-dichlorobenzoate.

A large number of studies have been conducted on metabolism of PCBs by mammalian and avian species.[123,126,141-155] The major metabolites which have been identified are usually phenols and catechols. Hass et al.[146] demonstrated the qualitative and quantitative aspects of the metabolism of 4-chlorobiphenyl in bile were 4-chloro-4-hydroxy biphenyl (65%), 4-chloro-3′,4′-dihydroxy biphenyl (22%) with some minor metabolites. The major metabolites of 4,4′-dichlorobiphenyl involved a chlorine shift to yield 3,4′-dichloro-4-hydroxy biphenyl (91%) and trace of 4,4′-dichloro-3-hydroxy biphenyl. Trihydroxy compounds,[154] methylated catechols,[146] and sulfur-containing metabolites[156-158] have also been identified, and in addition, dechlorinated metabolites have been observed.[148] The hydroxylated chlorobiphenyls which have been identified in animals suggest that the oxidation proceeds via arene oxide intermediates,[92,93] and this is supported by the isolation of *trans*-3,4-dihydro-3,4-dihydroxy-2,5,2′,5′-tetrachlorobiphenyl as a rabbit urinary metabolite of 2,5,2′,5′-tetrachlorobiphenyl.[126] Direct hydroxylation is also likely to occur in rats in the metabolism of PCBs.[155,159] However, ring cleavage which is common in bacteria has not been reported in mammalian and avian species.

VI. FATE OF PCBs IN THE ENVIRONMENT

Ballshmiter et al.[69] reported that some penta-, hexa-, and octachlorobiphenyls were found in marine fish liver in the same ratios as measured in the commercial PCB mixtures, Aroclor® 1254 and Aroclor® 1260. On the other hand, the decline of the relative content of 2,3,4,6,2′,5′-hexachlorobiphenyl, a major component in Aroclor® 1254 and 1260, down to about 20% of its calculated original input excludes recent pollution. A decline of the same extent is observed for the 2,3,4,2′,3′-penta, the 2,3,4,5,2′,3′-hexa and 2,3,4,5,6,2′,3′-heptachlorobiphenyls. All these PCBs have a nonrecalcitrant substitution pattern in one ring. Using the above constant ratio of PCB components, the original PCB input into the environment may be calculated.

The process of degradation of PCBs is of outstanding concern for the evaluation of those compounds in the environment. It has been shown by a number of workers[40-50] that PCBs decompose by pure cultures of microorganisms and naturally occurring microbial populations. The PCB-resistant or PCB-degrading bacteria seem to be ubiquitously distributed in the environment. Sayler et al.[70] found that bacteria capable of growth on media containing Aroclor® 1254 as the primary carbon source were recovered from all water and sediment samples collected from estuarine and marine environments. Fritz and Su[106] isolated 20 PCB-degrading bacteria representing 5 genera from river sediment and sewage effluent. Shiaris et al.[160] demonstrated that naturally occurring microorganisms are capable of significant PCB degradation in the aquatic environment. It is known that gene coding for degradation of some hydrocarbons and chlorinated pesticides such as toluene, xylene, salicylate, octane,[161,162] and 2,4-dichlorophenoxy acetic acid[163] occurs naturally in the form of transmissible or nontransmissible plasmids. A plasmid DNA coding for 4-chlorobiphenyl degradation has recently been found in a *Klebsiella* sp.[164] It now appears that plasmids specifying the bioconversion of PCBs may be able to evolve in soil and aquatic bacteria.

Some factors seem to affect inhibition in microbial degradation of PCBs in the environment. Tulp et al.[165] postulated that the metabolism of 4,4′-dichlorobiphenyl did not occur in mixed cultures of bacteria, if alternative carbon sources such as glucose, glycerol, and peptone, were used. PCBs have been detected in the sewage of wastewater treatment plants,[17,18] which is an indication that the activated sludge process does not completely remove PCBs, and supporting evidence for this has been provided by model laboratory experiments.[166] PCB is stable in soil or sediment. Niesen[167] reported that the microflora of soil were not able to attack Aroclor® 1254 under favorable conditions. Fries[168] found that chromatograms of Aroclor® 1254, which had been incubated in silage that underwent normal anaerobic fermentation for several months were identical to chromatograms of standard Aroclor® 1254. Carey and Harvey[49] observed no metabolism in an anaerobic marine mud incubated with a chlorophenyl isomer for 6 weeks. These results imply that in the large areas of anoxic sediments found in marshes, estuarines, and basins, PCBs will remain unaltered biologically or chemically. These anaerobic environments may then serve as long-term sinks for PCBs.

In order to fairly evaluate the persistence of PCBs it must be understood that PCBs are not a single entity, but complex mixtures made up of many entities which may undergo biological degradation at different rates. The chlorination of the mixtures is the most significant factor in the relative degradability of PCBs. A possible way to reduce the longevity of PCBs is to introduce or change substituents on the molecule to favor microbial attack. Laboratory studies of PCB metabolism include work on such diverse systems as rats,[141-146] rabbits,[126,147-149] birds,[123] pigs,[159] and primates.[153-155] Most of these studies report metabolic production of hydroxylated derivatives of various chlorinated isomers. A summary of various laboratory studies of PCB metabolism is given by Hutzinger et al.[32] The natural occurrence of PCB metabolites in feces from seals and guillemots has been reported.[156,169] Several experiments indicate that the degradation of PCBs is possible in such aquatic animals as goldfish,[170,171] rainbow trout,[172] frogs,[173] and marine annelids,[174] although the extent of metabolic degradation may be very low.

Another important degradation process of PCBs in the environment is photolysis by sunlight. It has been shown by a number of workers[175-181] that PCBs decomposed relatively easily on exposure to UV light in the laboratory. In both aqueous[175,176] and organic media,[177] PCBs react mainly by stepwise dechlorination (replacement of Cl by H). Substitution of Cl by OH seems to be a minor pathway, even in aqueous media. Thus photolysis converts highly chlorinated PCBs into less chlorinated ones. Bunce et al.[182] predict that a significant percentage of PCB molecules will lose at least one chlorine atom per year through photolysis. For shallow waters (e.g., Lake Erie, ≈20 m) they expect that up to 5% of lightly chlorinated PCBs might lose a chlorine atom each year, but at least one chlorine should be lost from every highly chlorinated molecule annually. Ruzo et al.[179,183] suggest that highly chlorinated PCBs are the most photolabile, and PCBs with ortho-chlorine substituents are photodecomposed more efficiently than those without. An example of this behavior is shown by 2,3,5,6-tetrachlorobiphenyl which alters <10% conversion when afforded equal amounts of 2,3,5-trichlorobiphenyl and secondary 3,5-dichlorobiphenyl.[183] These results are in sharp contrast to microbial degradation of PCBs,[117,119] in which degradation decreased as chlorine substitution increased, and PCBs containing two chlorines on either the ortho position of a single ring (i.e., 2,6) or both rings (i.e., 2,2′-) showed very poor degradability. If photolysis and microbial degradation are well combined in the environment, fast and efficient decomposition of PCBs with both high and low chlorination will be expected.

The decline of PCBs in the environment should be attributed to their dispersal by apparent dilution of dissolved PCBs and evaporation into the atmosphere, association of PCBs with sedimenting particles, and biological and chemical degradation. Harvey et al.[72,184] reported that in 1971 and 1972 average surface concentration of PCBs in the open North Atlantic Ocean was about 30 ng/ℓ.[184] By April of 1973 the concentrations had been reduced to 1 ng/ℓ;[72] 5 months later it was found that the decline had leveled at 0.8 ng/ℓ. However, these results have been challenged on technical grounds.[185] Unfortunately, there is relatively little information available on how much PCBs actually are declining in the environment; however, it is quite certain that PCBs in the environment are mainly degrading by both weathering and microbial processes.

ACKNOWLEDGMENTS

I express my thanks to Dr. Fumio Matsumura, of the Pesticide Research Center of the University of Michigan, Dr. Kenzo Tonomura of Osaka Prefectural University, Dr. Noboru Tomizuka and Dr. Akira Kamibayashi of the Fermentation Research Institute for their suggestions and helpful discussion.

REFERENCES

1. **Jensen, S.**, Report of a new chemical hazard, *New Sci.*, 32, 612, 1966.
2. **Risebrough, R. W., Reiche, P., Peakall, D. B., Herman, S. G., and Kirven, M. N.**, Polychlorinated biphenyls in the global ecosystem, *Nature (London)*, 220, 1098, 1968.
3. **Zikto, V., Hutzinger, O., and Safe, S.**, Retention times and electron-capture detection responses of some individual chlorobiphenyls, *Bull. Environ. Contam. Toxicol.*, 6, 160, 1971.
4. Monsanto Tech. Bull. G/PL-306, Aroclor Plasticizers.
5. **Fishbein, L.**, Chromatographic and biological aspects of polychlorinated biphenyls, *J. Chromatogr.*, 68, 345, 1972.
6. Polychlorinated Biphenyls in the Environment, U.S. Department of Commerce, Washington, D.C., 1972.
7. **Holmes, D. C., Simmons, J. H., and Tatton, J. O'G.**, Chlorinated hydrocarbons in British wildlife, *Nature (London)*, 216, 227, 1967.
8. **Jensen, S. and Widmark, G.**, Unintended Residue in the Environment, O.E.C.D. Rep., 1967.
9. **Holden, A. V. and Marsden, K.**, Organochlorine pesticides in seals and porpoises, *Nature (London)*, 216, 1274, 1967.
10. **Anderson, D. W., Hickey, J. J., Risebrough, R. W., Hughes, D. F., and Christensen, R. E.**, Significance of chlorinated hydrocarbon residues to breeding pelicans and cormorants, *Can. Field Nat.*, 83, 89, 1969.
11. **Koeman, J. H., Ten Noever De Brauw, M. C., and De Vos, R. H.**, Chlorinated biphenyls in fish, mussels and birds from the River Rhine and the Netherland coastal area, *Nature (London)*, 221, 1126, 1969.
12. **Jensen, S., Johnels, A. G., Olsson, M., and Otterlind, G.**, DDT and PCB in marine animals from Swedish waters, *Nature (London)*, 224, 247, 1969.
13. **Reynolds, L. M.**, Pesticide residue analysis in the presence of PCBs (review), *Residue Rev.*, 34, 27, 1971.
14. **Frank, R., Holdrinet, M., Braun, H. E., Dodge, D. P., and Sprargler, G. E.**, Residues of organochlorine insecticides and polychlorinated biphenyls in fish from Lakes Huron and Superior, Canada-1968—76, *Pest. Monit. J.*, 12, 60, 1978.
15. **Zikto, V., Hutzinger, O., and Choi, P. M. K.**, Contamination of the Bay of Fundy — Gulf of Marine area with PCBs, PCTs, Chlorinated DBF, and DBD, *Environ. Health Perspect.*, 1, 47, 1972.
16. **Veith, G. D.**, Baseline concentrations of polychlorinated biphenyls and DDT in Lake Michigan fish, 1971, *Pest. Monit. J.*, 9, 21, 1975.
17. **Lawrence, J. and Tosine, H. M.**, Polychlorinated biphenyl concentration in sewage and sludges of some waste treatment plants in southern Ontario, *Bull. Environ. Contam. Toxicol.*, 17, 49, 1977.
18. **Heesen, T. C. and Yong, D. R.**, Halogenated hydrocarbons in waste waters: knowns and unknowns, in Annual Report, Coastal Water Research Project, El Segundo, Calif., 1977, 33.
19. **Williams, D. T. and Benoit, F. M.**, The determination of polychlorinated biphenyls in selected household products, *Bull. Environ. Contam. Toxicol.*, 21, 179, 1979.
20. **Billings, W. N., Bidleman, T. F., and Varnberg, W. B.**, Movement of PCB from a contaminated reservoir into a drinking water supply, *Bull. Environ. Contam. Toxicol.*, 19, 215, 1978.
21. **Westöö, G. and Noren, K.**, Level of organochlorine pesticides and polychlorinated biphenyls in margarine, vegetable oils, and some foods of animal origin on the Swedish market in 1967—1969, *Var Foeda*, 22, 9, 1970.
22. **Highland, J.**, PCBs in food. What is acceptable? *Environment*, 18, 12, 1976.

23. **Westöö, G. and Noren, K.**, Levels of organochlorine pesticides and polychlorinated biphenyls in fish caught in Swedish water areas or kept for sale in Sweden, 1967—1970, *Var Foeda,* 22, 93, 1970.
24. **Hagel, P. and Tuinstra, G. M. Th.**, Trends in PCB concentration in Dutch coastal and inland fishery products 1972—1976. *Bull. Environ. Contam. Toxicol.,* 19, 671, 1978.
25. **Biros, F. J., Walker, A. C., and Medbury, A.**, Polychlorinated biphenyls in human adipose tissue, *Bull. Environ. Contam. Toxicol.,* 5, 317, 1970.
26. **Curley, A., Burse, V. W., Jenning, R. W., Villanue, E. C., Tomatis, L., and Akazaki, K.**, Chlorinated hydrocarbon pesticides and related compounds in adipose tissue from people of Japan, *Nature (London),* 242, 338, 1973.
27. **Acker, L. and Schulte, E.**, Occurrence of chlorinated biphenyls and hexachlorobenzene together with chlorinated insecticides in human milk and adipose tissue, *Naturwissenschaften,* 57, 497, 1970.
28. **Savage, E. P., Tessair, J. D., Malberg, J. W., Wheeler, H. W., and Bagby, J. R.**, Organochlorine pesticide residues and polychlorinated biphenyls in human milk, Colorado — 1971—1972, *Pest. Monit. J.,* 7, 15, 1973.
29. **Yoshida, S. and Nakamura, A.**, Residual status after parturition of methylsulfone metabolites of polychlorinated biphenyls in the breast milk of a former employee in a capacitor factory, *Bull. Environ. Contam. Toxicol.,* 21, 111, 1979.
30. **Yong, D. R., Ehrlich, D. M., and Heessen, T. C.**, Sediments as sources of DDT and PCB, *Mar. Pollut. Bull.,* 8, 254, 1977.
31. **Peakall, D. B. and Lincer, J. L.**, Polychlorinated biphenyls: another long-life, widespread chemical in the environment, *Biol. Sci.,* 20, 958, 1970.
32. **Hutzinger, O., Safe, S., and Zikto, V.**, *The Chemistry of PCBs,* CRC Press, Boca Raton, Fla., 1974.
33. **Veith, G. D. and Lee, G. F.**, A review of chlorinated biphenyl contamination in natural water, *Water Res.,* 4, 265, 1970.
34. WHO Task Group, Environmental Health Criteria for Polychlorinated Biphenyls and Terphenyls, World Health Organization, Geneva, 1975.
35. **Kimbrough, R. D.**, The toxicity of polychlorinated polycyclic compounds and related chemicals, *CRC Crit. Rev. Toxicol.,* 2, 445, 1975.
36. **Peakall, D. B. and Risebrough, R. W.**, PCBs and their environmental effects, *CRC Crit. Rev. Environ. Control,* 5, 496, 1975.
37. **Nelson, N.**, Polychlorinated biphenyls — environmental impact, *Environ. Res.,* 5, 249, 1972.
38. **Fishbein, L.**, Toxicity of chlorinated biphenyls, *Ann. Rev. Pharmacol.,* 14, 139, 1974.
39. **Kimbrough, R., Buckley, J., Fishbein, L., Flamm, G., Kasza, L., Marcus, W., Shibko, S., and Teske, R.**, Animal toxicology, in Final report of the subcommittee on the health effects of polychlorinated biphenyls and polybrominated biphenyls of DHEW committee to coordinate toxicology and related programs, *Environ. Health Perspect.,* 24, 173, 1978.
40. **Ahmed, M. and Focht, D. D.**, Degradation of polychlorinated biphenyls by two species of *Achromobacter, Can. J. Microbiol.,* 19, 47, 1973.
41. **Wallnöfer, P. R., Engelhardt, G., Safe, S., and Hutzinger, O.**, Microbial hydroxylation of 4-chlorobiphenyl and 4,4′-dichlorobiphenyl, *Chemosphere,* 2, 69, 1973.
42. **Wong, P. T. S. and Kaiser, K. L. E.**, Bacterial degradation of polychlorinated biphenyls. II. Rate studies, *Bull. Environ. Contam. Toxicol.,* 13, 249, 1975.
43. **Tucker, E. S., Saeger, V. W., and Hicks, O.**, Activated sludge primary biodegradation of polychlorinated biphenyls, *Bull. Environ. Contam. Toxicol.,* 14, 705, 1975.
44. **Baxter, R. A., Gilbert, P. E., Lidgett, R. A., Mainprize, J. H., and Vodden, H. A.**, The degradation of polychlorinated biphenyls by microorganisms, *Sci. Total Environ.,* 4, 53, 1975.
45. **Mihashi, K., Hashinaga, T., Dazai, M., Ito, J., and Misono, T.**, Biodegradability of polychlorobiphenyl, *Kogyo Gijutsuin Biseibutsu Kogyo Gijutsu Kenkyusho Kenkyu Hokoku,* 46, 65, 1975.
46. **Furukawa, K. and Matsumura, F.**, Microbial metabolism of polychlorinated biphenyls: studies on relative degradability of polychlorinated biphenyl components by *Alcaligenes* sp., *J. Agric. Food Chem.,* 24, 251, 1976.
47. **Ballshmiter, K., Unglert, K. C., and Neu, H. T.**, Abbau von chlorierten Aromaten: Mikrobiologischer Abbau der polychlorierte Biphenyle (PCB). III. Chlorierte Benzoesäuren als Metabolite der PCB, *Chemosphere,* 1, 51, 1977.
48. **Sayler, G. S., Shou, M., and Colwell, R. R.**, Growth of an estuarine *Pseudomonas* sp. on polychlorinated biphenyls, *Microb. Ecol.,* 3, 241, 1977.
49. **Carey, A. E. and Harvey, G. R.**, Metabolism of polychlorinated biphenyls by marine bacteria, *Bull. Environ. Contam. Toxicol.,* 20, 527, 1978.
50. **Clark, R. R., Chian, E. S. K., and Griffin, R. A.**, Degradation of polychlorinated biphenyls by mixed microbial cultures, *Appl. Environ. Microbiol.,* 37, 680, 1979.
51. **Miure, J. P., Hicks, O., Kaley, P. G., and Saeger, V. W.**, Characterization of polychlorinated biphenyls, in *Conf. Proc. Natl. Conf. Polychlorinated Biphenyls,* 1976, 84.

52. **Munson, T. O., Ela, D. K., and Rutleadge, C.**, Upper Bay Survey, Final Report to the Maryland Dept. of Natural Resources, Vol. 2, Westinghouse Electric Corp., Annapolis, Md., 1975.
53. **Oloffs, P. C., Albright, L. J., and Szeto, S. Y.** Fate and behavior of five chlorinated hydrocarbons in three natural waters, *Can. J. Microbiol.*, 18, 1393, 1972.
54. **Paris, D. F., Steen, W. C., and Baughman, G. L.**, Role of physico-chemical properties of Aroclors 1016 and 1242 in determining their fate and transport in aquatic environments, *Chemosphere*, 4, 319, 1978.
55. **Weil, L., Dure, G., and Quentri, K.**, Wasserlöslichkeit von insektiezieden chlorierten Kohlenwasserstoffen und polychlorierten Biphenylen in Hinblick auf eine Gewasserbelastung mit diesen Stofen, *Wasser Abwasser*, 7, 169, 1974.
56. **Cooley, N. R., Keltner, J. M., and Forester, J.**, The polychlorinated biphenyls, Aroclors 1248 and 1260: effect on and accumulation by *Tetrahymena pyriformis*, *J. Protozool.*, 20, 443, 1973.
57. **Urey, J. C., Kricher, J. C., and Boylan, J. M.**, Bioaccumulation of four pure PCB isomers by *Chlorella pyrenoidosa*, *Bull. Environ. Contam. Toxicol.*, 16, 81, 1976.
58. **Johnson, B. T. and Kennedy, J. O.**, Biomagnification of p,p'-DDT and methoxychlor by bacteria, *Appl. Microbiol.*, 26, 66, 1973.
59. **Sayler, G. S. and Colwell, R. R.**, Partioning of mercury and polychlorinated biphenyls (PCB) by oil, water, and suspended sediment, *Environ. Sci. Technol.*, 10, 1142, 1976.
60. **Eder, G.**, Polychlorinated biphenyls and compounds of the DDT group in sediments of the Central North Sea and Norwegian depression, *Chemosphere*, 2, 101, 1976.
61. **Paris, D. F. and Lewis, D. L.**, Accumulation of methoxychlor by microorganisms isolated from aqueous systems, *Bull. Environ. Contam. Toxicol.*, 15, 24, 1976.
62. **Paris, D. F., Lewis, D. L., and Barnett, J. T.**, Bioconcentration of toxaphene by microorganisms, *Bull. Environ. Contam. Toxicol.*, 17, 564, 1977.
63. **Rice, C. P. and Sikka, H. C.**, Uptake and metabolism of DDT by six species of marine algae, *J. Agric. Food Chem.*, 21, 148, 1973.
64. **Södergren, A.**, Chlorinated hydrocarbon residues in airborne fallout, *Nature (London)*, 236, 395, 1972.
65. **Hammond, A. L.**, Chemical pollution: polychlorinated biphenyls, *Science*, 175, 155, 1972.
66. **Södergren, A.**, Transport, distribution, and degradation of DDT and PCB in a south Swedish lake ecosystem, *Vatten*, 2, 90, 1973.
67. **Greichus, Y. A., Greichus, A., Draayer, H. A., and Marshall, B.**, Insecticides, polychlorinated biphenyls and metals in African lake ecosystems. II. Lake McIlwaine, Rhodesia, *Bull. Environ. Contam. Toxicol.*, 19, 444, 1978.
68. **Ernst, W., Goerke, H., Eder, G., and Schaefer, R. G.**, Residue of chlorinated hydrocarbons in marine organisms in relation to size and ecological parameters. I. PCB, DDT, DDE, and DDD in fishes and molluscs from the English Channel, *Bull. Environ. Contam. Toxicol.*, 15, 55, 1976.
69. **Ballshmiter, K., Zell, M., and Neu, H. J.** Resistance of PCBs in the ecosphere: Will some PCB-components "never" degrade? *Chemosphere*, 2, 173, 1978.
70. **Sayler, G. S., Thomas, R., and Colwell, R. R.**, Polychlorinated biphenyl (PCB) degrading bacteria and PCB in estuarine and marine environment, *Estuarine Coastal Mar. Sci.*, 6, 553, 1978.
71. **Elder, D. L. and Villeneuve, J. P.**, Polychlorinated biphenyls in the Mediterranean Sea, *Mar. Pollut. Bull.*, 8, 19, 1977.
72. **Harvey, G. R., Steinhauser, W. G., and Mikeas, H. P.**, Decline of PCB concentrations in North Atlantic surface water, *Nature (London)*, 252, 387, 1974.
73. **Tatsukawa, R. and Wakimoto, T.**, The fate of PCBs in the environment, in *Environmental Quality and Safety*, Coulson, F. and Korte, F., Eds., Thieme, Stuttgart, 1975, 188.
74. **Furukawa, K., Matsumura, F., and Tonomura, K.**, *Alcaligenes* and *Acinetobacter* strains capable of degrading polychlorinated biphenyls, *Agric. Biol. Chem.*, 42, 542, 1978.
75. **Murado, M. A., Tejedor, M. C., and Baluja, G.**, Interactions between polychlorinated biphenyls (PCBs) and soil microfungi. Effect of Aroclor 1254 and other PCBs on *Aspergillus flavus* cultures, *Bull. Environ. Contam. Toxicol.*, 15, 768, 1976.
76. **Choi, P. S., Nack, H., and Flinn, J. E.**, Distribution of polychlorinated biphenyls in an aerated biological oxidation waste water treatment system, *Bull. Environ. Contam. Toxicol.*, 11, 12, 1974.
77. **Sanders, H. O. and Chandler, J. H.**, Biological magnification of a polychlorinated biphenyl (Aroclor 1254) from water by aquatic invertebrates, *Bull. Environ. Contam. Toxicol.*, 7, 257, 1972.
78. **Metcalf, R. L., Sanborn, J. R., Lu, P. Y., and Nye, D.**, Laboratory model ecosystems of the degradation and fate of radiolabeled tri-, tetra-, and pentachlorobiphenyl compared with DDE, *Arch. Environ. Contam. Toxicol.*, 3, 151, 1975.
79. **Keil, J. E., Priester, L. E., and Sandifer, S. H.**, Polychlorinated biphenyl (Aroclor 1242): effect of uptake on growth, nucleic acids, and chlorophyll of a marine diatom, *Bull. Environ. Contam. Toxicol.*, 6, 156, 1971.

80. **Mosser, J. L., Fisher, N. S., Teng, T., and Wuster, C. F.**, Polychlorinated biphenyls: toxicity to certain phytoplankton, *Science*, 175, 191, 1972.
81. **Glooschenko, V. and Glooschenko, W.**, Effect of polychlorinated biphenyl compounds on growth of Great Lake phytoplankton, *Can. J. Bot.*, 53, 653, 1975.
82. **Moore, S. A. and Harriss, R. C.**, Effects of polychlorinated biphenyl on marine phytoplankton communities, *Nature (London)*, 240, 356, 1972.
83. **Powers, C. D., Rowland, R. G., O'Connors, H. B., and Wurster, C. F.**, Response to polychlorinated biphenyls of marine phytoplankton isolates culture under natural conditions, *Appl. Environ. Microbiol.*, 34, 760, 1977.
84. **Mosser, J. L., Fisher, N. S., and Wurster, C. F.**, Polychlorinated biphenyls and DDT alter species composition in mixed cultures of algae, *Science*, 176, 533, 1972.
85. **Fisher, N. S., Carpenter, E. J., Remsen, C. C., and Wurster, C. F.**, Effects of PCB on interspecific composition in natural and gnotobiotic phytoplankton communities in continuous cultures, *Microb. Ecol.*, 1, 39, 1974.
86. **Ewald, W. G., French, J. E., and Champ, M. A.**, Toxicity of polychlorinated biphenyls (PCBs) to *Euglena gracilis*: cell population growth, carbon fixation, chlorophyll level, oxygen consumption, and protein and nucleic acid synthesis, *Bull. Environ. Contam. Toxicol.*, 16, 71, 1976.
87. **Hawes, M. L., Kricker, J. C., and Urey, J. C.**, The effects of various Aroclor fractions on the population growth of *Chlorella pyrenoidosa*, *Bull. Environ. Contam. Toxicol.*, 15, 14, 1976.
88. **Zullei, N. and Benecke, G.**, Application of a new bioassay to screen the toxicity of polychlorinated biphenyls on blue-green algae, *Bull. Environ. Contam. Toxicol.*, 20, 786, 1978.
89. **Craigie, J. S. and Hutzinger, O.**, Effects of commercial chlorobiphenyls on the growth of seven species of marine phytoplankton, *Chemosphere*, 3, 139, 1975.
90. **Lichtenstein, E. P., Schulz, K. R., Fuhremann, T. W., and Liang, T. T.**, Biological interaction between plasticizers and insecticides, *J. Ecol. Entomol.*, 62, 761, 1969.
91. **Yoshimura, H., Yamamoto, H., and Saeki, S.**, Metabolic studies of polychlorinated biphenyls. II. Metabolic fate of 2,4,3',4'-tetrachlorobiphenyl in rats, *Chem. Pharm. Bull.*, 21, 2231, 1973.
92. **Daly, J. W., Jerina, D. M., and Witkop, B.**, Arene oxides and the NIH shift: the metabolism, toxicity, carcinogenicity of aromatic compounds, *Experientia*, 28, 1129, 1972.
93. **Jerina, D. M. and Daly, J. W.**, Arene oxides: a new aspect of drug metabolism, *Science*, 185, 573, 1974.
94. **Heidelberger, C.**, Chemical carcinogenesis, *Annu. Rev. Biochem.*, 44, 79, 1975.
95. **Furukawa, K., Tomizuka, N., and Kamibayashi, A.**, Bacterial Degradation of Commercial PCB Mixtures (Kaneclors) and Their Metabolites, *J. Pharm. Dyn.*, 3, S-4, 1980.
96. **Bowes, G. W. and Mulvihill, M. J.**, Identification of chlorinated dibenzofurans in American polychlorinated biphenyls, *Nature (London)*, 256, 305, 1975.
97. **Curley, A., Burse, V. W., Jennings, R. W., Villaneueva, E. C., and Kimbrough, R. D.**, Evidence of tetrachlorodibenzofuran (TCDF) in Aroclor 1254, and the urine of rats following dietary exposure to Aroclor 1254, *Bull. Environ. Contam. Toxicol.*, 14, 2, 1975.
98. **Bowes, G. W., Mulvihill, M. J., DeCamp, M. R., and Kende, A. S.**, Gas chromatographic characteristics of authentic chlorinated dibenzofurans; identification of two isomers in American and Japanese polychlorinated biphenyls, *J. Agric. Food Chem.*, 23, 1222, 1975.
99. **Crosby, D. G. and Moilanen, K. W.**, Photodecomposition of chlorinated biphenyls and dibenzofurans, *Bull. Environ. Contam. Toxicol.*, 10, 372, 1973.
100. **Morita, M., Nakagawa, J., and Rappe, C.**, Polychlorinated dibenzofuran (PCDF) formation from PCB mixture by heat and oxygen, *Bull. Environ. Contam. Toxicol.*, 19, 665, 1978.
101. **Lunt, D. and Evans, C.**, The microbial metabolism of biphenyl, *Biochem. J.*, 118, 54, 1970.
102. **Catelani, D., Sorlini, C., and Treccani, V.**, The metabolism of biphenyl by *Pseudomonas putida*, *Experientia*, 27, 1173, 1971.
103. **Catelani, D., Colombi, A., Sorlini, C., and Treccani, V.**, Metabolism of biphenyl. 2-Hydroxy-6-oxo-6-phenylhexa-2,4-dienoate: the meta-cleavage product from 2,3-dihydroxybiphenyl by *Pseudomonas putida*, *Biochem. J.*, 134, 1063, 1973.
104. **Ohmori, T., Ikai, T., Minoda, Y., and Yamada, K.**, Utilization of polyphenyl and polyphenyl-related compounds by microorganisms, *Agric. Biol. Chem.*, 37, 1599, 1973.
105. **Keil, J. E., Sandifer, S. H., Graber, C. D., and Priester, L. E.**, DDT and polychlorinated biphenyl (Aroclor 1242). Effect of uptake on *E. coli* growth, *Water Res.*, 6, 837, 1972.
106. **Fritz, A. K. and Su, T. M.**, The Isolation and Characterization of Bacteria that Degrade Polychlorinated Biphenyls (PCB), Abstracts of the ASM annual meeting — 1978, 203.
107. **Bourquin, A. W. and Cassidy, S.**, Effect of polychlorinated biphenyl formations on the growth of estuaraine bacteria, *Appl. Microbiol.*, 29, 125, 1975.
108. **Blakemore, R. P. and Carey, A. E.**, Effect of polychlorinated biphenyls on growth and respiration of heterotrophic marine bacteria, *Appl. Environ. Microbiol.*, 35, 323, 1978.

109. **Wyndham, C., Devenish, J., and Safe, S.**, The *in vitro* metabolism, macromolecular binding and bacterial mutagenicity of 4-chlorobiphenyl, a model PCB substrate, *Res. Commun. Chem. Pathol. Pharmacol.*, 15, 563, 1976.
110. **Smith, R. V. and Rosazza, J. P.**, Microbial models of mammalian metabolism. Aromatic hydroxylation, *Arch. Biochem. Biophys.*, 161, 551, 1974.
111. **Dodge, R. H., Cerniglia, C. E., and Gibson, D. H.**, Fungal metabolism of biphenyl, *Biochem. J.*, 178, 223, 1979.
112. **Tejedor, M. C., Murado, M. A., and Balija, G.**, Oxidative metabolism in *Saccharomyces cerevisiae* as affected by polychlorinated biphenyls, *Bull. Environ. Contam. Toxicol.*, 22, 439, 1979.
113. **Wiseman, A., Gondal, J. A., and Sims, P.**, 4-Hydroxylation of biphenyl by yeast containing cytochrome P-450: radiation and thermal stability, comparisons with liver enzyme (oxidized and reduced forms), *Biochem. Soc. Trans.*, 3, 278, 1975.
114. **Knackmus, H. J. and Reineke, W.**, Der Einfluss von Chlor-substituenten auf die Oxygenierung von Benzoat durch *Alcaligenes eutrophus* B9, *Chemosphere*, 2, 225, 1973.
115. **Peakall, D. B. and Lincer, J. L.**, Polychlorinated biphenyls — another long life widespread chemical in the environment, *BioScience*, 20, 958, 1970.
116. **Horvath, R. S.**, Microbial cometabolism and the degradation of organic compounds in nature, *Bacteriol. Rev.*, 36, 146, 1972.
117. **Furukawa, K., Tonomura, K., and Kamibayashi, A.**, Effect of chlorine substitution on the biodegradability of polychlorinated biphenyls, *Appl. Environ. Microbiol.*, 35, 223, 1978.
118. **Widmark, G.**, The OECD Study of Analysis of PCB, Report from the Institute of Analytical Chemistry, Stockholm, 1968.
119. **Furukawa, K., Tomizuka, N., and Kamibayashi, A.**, Effect of chlorine substitution on the bacterial metabolism of various polychlorinated biphenyls, *Appl. Environ. Microbiol.*, 38, 301, 1979.
120. **Melves, B. and Brandt, I.**, The distribution and metabolism of labeled polychlorinated biphenyls in mice and quails, presented at PCB Conference. III. National Swedish Environment Board Publication 4E, 1973, 87.
121. **Berlin, M., Gage, J., and Holn, S.**, The distribution and metabolism of 2,4,5,2′,5′-pentachlorobiphenyl, *Arch. Environ. Health*, 30, 141, 1975.
122. **Matthews, H., Fries, G., Gardner, A., Garthoff, L., Goldstein, J., Ku, Y., and Moore, J.**, Metabolism and biological toxicity of PCBs and PBBs, *Environ. Health Perspect.*, 24, 147, 1978.
123. **Hutzinger, O., Nash, D. M., Safe, S., DeFreitas, A. S. W., Norström, R. J., Wildish, D. J., and Zikto, V.**, Polychlorinated biphenyls: metabolic behavior of pure isomers in pigeons, rats, and brook trout, *Science*, 178, 313, 1972.
124. **Matthews, H. B. and Anderson, M. W.**, Effect of chlorination on the distribution and excretion of polychlorinated biphenyls, *Drug Metab. Dispos.*, 3, 371, 1975.
125. **Lucier, G. W., McDaniel, O. S., Schiller, C. M., and Matthews, H. B.**, Structural requirements for the accumulation of chlorinated biphenyl metabolites in the fetal rat intestine, *Drug Metab. Dispos.*, 6, 584, 1978.
126. **Gardner, A. M., Chen, J. T., Roach, J. A. G., and Ragelis, E. P.**, Polychlorinated biphenyls: hydroxylated urinary metabolites of 2,5,2′,5′-tetrachlorobiphenyl identified in rabbits, *Biochem. Biophys. Res. Commun.*, 55, 1377, 1973.
127. **DeFreitas, A. S. and Norström, R. J.**, Turnover and metabolism of PCBs in relation to their chemical structure and movement of lipids in the pigeon, *Can. J. Physiol.*, 52, 1080, 1974.
128. **Furukawa, K., Tonomura, K., and Kamibayashi, A.**, Metabolism of 2,4,4′-trichlorobiphenyl by *Acinetobacter* sp. P6, *Agric. Biol. Chem.*, 43, 1577, 1979.
129. **Gibson, D. T., Cardini, G. E., Maseles, F. C., and Kalio, R. E.**, Incorporation of oxygen-18 into benzene by *Pseudomonas putida*, *Biochemistry*, 9, 1631, 1970.
130. **Reiner, A. M. and Hegeman, G. D.**, Metabolism of benzoic acid by bacteria. Accumulation of (−)3,5-cyclohexadiene-1,2-diol-1-carboxylic acid by a mutant strain of *Alcaligenes eutrophus*, *Biochemistry*, 10, 2530, 1971.
131. **Jerina, D. M., Daly, D. W., Jeffery, A. M., and Gibson, D. T.**, *cis*-1,2-Dihydroxy-1,2-dihydronaphthalene: a bacterial metabolite from naphthalene, *Arch. Biochem. Biophys.*, 142, 394, 1971.
132. **Gibson, D. T., Roberts, R. L., Wells, M. C., and Kobal, U. M.**, Oxidation of biphenyl by a *Beijerinckia* species, *Biochem. Biophys. Res. Commun.*, 50, 211, 1973.
133. **Kaiser, K. L. E. and Wong, P. T. S.**, Bacterial degradation of polychlorinated biphenyls. I. Identification of some metabolic products from Aroclor 1242, *Bull. Environ. Contam. Toxicol.*, 11, 291, 1974.
134. **Horvath, R. S. and Alexander, M.**, Cometabolism of *m*-chlorobenzoate by an *Arthrobacter*, *Appl. Microbiol.*, 20, 254, 1970.
135. **Ruisinger, S., Klages, U., and Lingens, F.**, Abbau der 4-Chlorobenzoesäure durch eine *Arthrobacter* Species, *Arch. Microbiol.*, 110, 253, 1976.

136. **Horvath, R. S.,** Cometabolism of the herbicide 2,3,6-trichlorobenzoate, *J. Agric. Food Chem.*, 19, 291, 1971.
137. **Dorn, E., Hellwigm, M., Reineke, W., and Knackmus, H. J.,** Isolation and characterization of a 3-chlorobenzoate degrading *Pseudomonad, Arch. Microbiol.*, 99, 61, 1974.
138. **Hartmann, J., Reineke, W., and Knackmuss, H. J.,** Metabolism of 3-chloro-, 4-chloro-, and 3,5-dichlorobenzoate by a *pseudomonad, Appl. Environ. Microbiol.*, 37, 421, 1979.
139. **Dewey, O. R., Lyndsay, R. V., and Hartley, G. S.,** Biological destruction of 2,3,6-trichlorobenzoic acid in soil, *Nature (London)*, 195, 1232, 1962.
140. **GiGeronimo, M. J., Nikaido, M., and Alexander, M.,** Utilization of chlorobenzoates by microbial populations in sewage, *Appl. Environ. Microbiol.*, 37, 619, 1979.
141. **Yoshimura, H. and Yamamoto, H.,** Metabolic studies on polychlorinated biphenyls. I. Metabolic fate of 3,4,3′,4′-tetrachlorobiphenyl in rats, *Chem. Pharm. Bull.*, 21, 1168, 1973.
142. **Goto, M., Sugiura, K., Hattori, M., Miyagawa, T., and Okamura, M.,** Metabolism of 2,3,5,6-tetrachlorobiphenyl-^{14}C and 2,3,4,5,6-pentachlorobiphenyl-^{14}C in the rat, *Chemosphere*, 5, 227, 1974.
143. **Safe, S., Hutzinger, O., and Ecobichon, D.,** Identification of 4-chloro-4′-hydroxybiphenyl and 4,4′-dichloro-3-hydroxy-biphenyl as metabolites of 4-chloro and 4,4′-dichlorobiphenyl fed to rats, *Experientia*, 30, 720, 1974.
144. **Jensen, S. and Sundström, G.,** Metabolic hydroxylation of a chlorobiphenyl containing only isolated unsubstituted positions-2,2′,4,4′,5,5′-hexachlorobiphenyl, *Nature (London)*, 251, 219, 1974.
145. **Greb, W., Klein, W., Coulston, F., Goldberg, L., and Korte, F.,** *In vitro* metabolism of polychlorinated biphenyls-^{14}C, *Bull. Environ. Contam. Toxicol.*, 13, 424, 1975.
146. **Hass, J. R., Jao, L. T., Wilson, N. K., and Matthew, H. B.,** Metabolism of 4-chlorobiphenyl and 4,4′-dichlorobiphenyl in the rat: qualitative and quantitative aspects, *J. Agric. Food Chem.*, 25, 1330, 1977.
147. **Block, W. D. and Cornish, H. H.,** Metabolism of biphenyl and 4-chlorobiphenyl in the rabbit, *J. Biol. Chem.*, 234, 3301, 1959.
148. **Hutzinger, O., Jamieson, W. D., Safe, S., Paulmann, L., and Ammon, R.,** Metabolic dechlorination of a highly chlorinated biphenyl in the rabbit. Rapid unambiguous identification of organochlorine metabolites by a high resolution mass spectrometric method, *Nature (London)*, 252, 698, 1974.
149. **Safe, S., Hutzinger, O., and Jones, D.,** The mechanism of chlorobiphenyl metabolism, *J. Agric. Food Chem.*, 23, 851, 1975.
150. **Safe, S., Ruzo, L. O., Jones, D., Platonow, N. S., and Hutzinger, O.,** The metabolism of 4-chlorobiphenyl in the pig, *Can. J. Physiol. Pharmacol.*, 53, 392, 1975.
151. **Safe, S., Platonow, N. S., and Hutzinger, O.,** The metabolism of chlorobiphenyls in the goat and cow, *J. Agric. Food Chem.*, 23, 259, 1975.
152. **Gardner, A. M., Gighter, H. F., and Roach, J. A. G.,** Excretion of hydroxylated polychlorinated biphenyl metabolites in cow's milk, *J. Assoc. Off. Anal. Chem.*, 59, 273, 1976.
153. **Greb, W., Klein, W., Coulston, F., Goldberg, L., and Korte, F.,** Excretion rates of pure di- and trichlorobiphenyl-^{14}C in the rhesus monkey, *Chemosphere*, 4, 143, 1973.
154. **Greb, W., Klein, W., Coulston, F., Goldberg, L., and Korte, F.,** Metabolism of lower polychlorinated biphenyls-^{14}C in the rhesus monkey, *Bull. Environ. Contam. Toxicol.*, 13, 471, 1975.
155. **Hsu, I. C., Van Miller, J. P., and Allen, J. R.,** Metabolic fate of ^{3}H-2,5,2′,5′-tetrachlorobiphenyl in infant nonhuman primates, *Bull. Environ. Contam. Toxicol.*, 14, 233, 1975.
156. **Jensen, S. and Jansson, B.,** Methyl sulfone metabolites of PCB and DDE, *Ambio*, 5, 257, 1976.
157. **Mio, T., Sumino, K., and Mizutani, T.,** Sulfur-containing metabolites of 2,5,2′,5′-tetrachlorobiphenyl, a major component of commercial PCBs, *Chem. Pharm. Bull.*, 24, 1958, 1976.
158. **Mizutani, T., Yamamoto, K., and Tajima, K.,** Sulfur-containing metabolites of chlorobiphenyl isomers, a comparative study, *J. Agric. Food Chem.*, 26, 862, 1978.
159. **Safe, S., Hutzinger, O., Ecobichon, D., and Grey, A. A.,** The metabolism of 4′-chloro-4-biphenylol in the rat, *Can. J. Biochem.*, 53, 415, 1975.
160. **Shiaris, M. P., Lund, L. C., and Sayler, G. S.,** Degradation of PCB by naturally occurring aquatic bacterial populations, in Abstracts of the ASM annual meeting — 1978, 195.
161. **Chakrabarty, A. M., Friello, D. A., and Bopp, L. H.** recA-dependent excision of a hydrocarbon degradative plasmid (TOL*) and its transposition onto antibiotic resistance plasmids, in *Plasmids and Other Extrachromosomal Genetic Elements*, Timmis, K. and Goebel, W., Eds., Max Plank Institute for Molecular Genetics, Berlin, 1978, 18.
162. **Chakrabarty, A. M., Friello, D. A., and Bopp, L. H.,** Transposition of plasmid DNA segments specifying hydrocarbon degradation and their expression in various microorganisms, *Proc. Natl. Acad. Sci. U.S.A.*, 75, 3109, 1978.
163. **Pemberton, J. M. and Fisher, P. R.,** 2,4-D plasmids and persistence, *Nature (London)*, 268, 732, 1977.

164. **Chakrabarty, A. M.**, Molecular mechanisms in the biodegradation of environmental pollutants, *ASM News*, 44, 687, 1978.
165. **Tulp, M. Th., Schmitz, R., and Hutzinger, O.**, The bacterial metabolism of 4,4′-dichlorobiphenyl, and its suppression by alternative carbon sources, *Chemosphere*, 1, 103, 1978.
166. **Herbst, E., Scheunert, I., Klein, W., and Korte, F.**, Fate of PCBs-^{14}C in sewage treatment — laboratory experiments with activated sludge, *Chemosphere*, 11, 725, 1977.
167. **Niesen, T. V.**, PCB i jordbunden. Stability of PCB in soil, *Statens Forsøgsvirksomhed i Plantekultur*, 1112, 533, 1973.
168. **Fries, G. F.**, Degradation of chlorinated hydrocarbons under anaerobic conditions, *Adv. Chem. Ser.*, 111, 256, 1972.
169. **Jansson, B., Jensen, S., Olsson, M., Renberg, L., Sundström, G., and Vaz, R.**, Identification by GC-MS of phenolic metabolites of PCB and p,p′-DDE isolated from Baltic Guillemot and seal, *Ambio*, 4, 93, 1975.
170. **Herbst, E., Weisgerber, I., Klein, W., and Korte, F.**, Beiträge zur ökologishen Chemie. CXVIII. Bilanz, Bioakumulierung und Unwandlung von 2,2′-Dichlorbiphenyl-^{14}C in Goldfishen, *Chemosphere*, 2, 127, 1976.
171. **Hinz, R. and Matsumura, F.**, Comparative metabolism of PCB isomers by three species of fish and rat, *Bull. Environ. Contam. Toxicol.*, 18, 631, 1977.
172. **Melancon, M. J., Jr. and Lech, J. J.**, Isolation and identification of a polar metabolite of tetrachlorobiphenyl from bile of rainbow trout exposed to ^{14}C-tetrachlorobiphenyl, *Bull. Environ. Contam. Toxicol.*, 15, 181, 1976.
173. **Safe, S., Jones, D., Kohli, J., Ruzo, L. O., Hutzinger, O., and Sundstrom, G.**, The metabolism of chlorinated aromatic pollutant by the frog, *Can. J. Zool.*, 54, 1818, 1976.
174. **Ernst, W., Goerke, H., and Weber, K.**, Fate of ^{14}C-labeled di-, tri-, and pentachlorobiphenyl in the marine annelid *Nereis virens*. II. Degradation and faecal elimination, *Chemosphere*, 9, 559, 1977.
175. **Safe, S. and Hutzinger, O.**, Polychlorinated biphenyls: Photolysis of 2,4,6,2′,4′,6′-hexachlorobiphenyl, *Nature (London)*, 232, 641, 1971.
176. **Hustert, K. and Korte, F.**, Beitrage zur okologishen Chemie. XXXVIII. Synthese polychlorierter Biphenyle und ihre Reaktionen bei UV-Bestrahlung, *Chemosphere*, 1, 7, 1972.
177. **Herring, J. L., Hannan, E. J., and Bills, D. D.**, UV irradiation of Aroclor 1254, *Bull. Environ. Contam. Toxicol.*, 8, 153, 1972.
178. **Ruzo, L. O., Zabik, M. J., and Schetz, R. D.**, Polychlorinated biphenyls: photolysis of 3,4,3′,4′-tetrachlorobiphenyl and 4,4′-dichlorobiphenyl in solution, *Bull. Environ. Contam. Toxicol.*, 8, 217, 1972.
179. **Ruzo, L. O., Zabik, M. J., and Schuetz, R. D.**, Photochemistry of bioactive compounds. Photochemical process of polychlorinated biphenyls, *J. Am. Chem. Soc.*, 96, 3809, 1974.
180. **Carey, J. H., Lawrence, J., and Tosine, H. M.**, Photodechlorination of PCBs in the presence of titanium dioxide in aqueous suspensions, *Bull. Environ. Contam. Toxicol.*, 16, 697, 1976.
181. **Bunce, N. J.**, Photolysis of 2-chlorobiphenyl in aqueous acetnitrile, *Chemosphere*, 8, 653, 1978.
182. **Bunce, N. J., Kumar, Y., and Brownlee, B. G.**, An assessment of the impact of solar degradation of polychlorinated biphenyls in the aquatic environment, *Chemosphere*, 2, 155, 1978.
183. **Ruzo, L. O., Safe, S., and Zabik, M. J.**, Photodecomposition of unsymmetrical polychlorobiphenyls, *J. Agric. Food Chem.*, 23, 594, 1975.
184. **Harvey, G. R., Steinhauer, W. G., and Teal, J. M.**, Polychlorobiphenyls in North Atlantic Ocean water, *Science*, 180, 643, 1973.
185. **Longhurst, A. R. and Radford, P. J.**, PCB concentrations in North Atlantic surface water, *Nature (London)*, 256, 239, 1975.

Chapter 3

CHEMICAL TRANSFORMATIONS OF AROMATIC HYDROCARBONS THAT SUPPORT THE GROWTH OF MICROORGANISMS

Douglas W. Ribbons and Richard W. Eaton

TABLE OF CONTENTS

I. INTRODUCTION

It is 75 years since Stormer[1] observed that microorganisms can use aromatic hydrocarbons as nutrients for assimilation. Some 40 years later Zobell[2] provided a very comprehensive review on the action of microorganisms on hydrocarbons. Unlike numerous later reviews of the last 3 decades on this topic there was little biochemical information available to discuss; instead Zobell elaborated on the natural and economic importance of microorganisms as agents for modifying hydrocarbons to useful products, as well as their role in the biodeterioration of several materials such as formulated natural and synthetic mixtures and polymers. Since then there have appeared numerous reviews[3-11] of the progress made in understanding the processes which microorganisms use, and acquire, for the transformations of aromatic hydrocarbons and their substituted derivatives; and it is becoming clear that the biochemical and genetic strategies evolved by "Nature's biodegraders" have been directed by physiochemical constraints of reaction mechanisms and specificities. Such conclusions have been largely due to concomitant studies made on the utilization of derivatives of aromatic hydrocarbons that are already partially oxidized, such as phenols and carboxylic acids. There are several excellent recent reviews that cover the metabolism of aromatic compounds in general,[12-17] and some, notably by Dagley, describe the unique fitness that some microbes have acquired for the oxidation of very reduced organic compounds.[14-16]

One of the very interesting and important observations made in the last few years by Cerniglia and Gibson is that some eukaryotic microbes, e.g., *Cunninghamella elegans*,[18] use mono-oxygenative enzymic mechanisms for the ring attack of aromatic hydrocarbons, as do mammals, to initiate detoxication reactions (Figure 1); hydration of the arene oxides gives *trans*-dihydrodiols. Dioxygenation initiates the degradation of aromatic hydrocarbon rings in a wide variety of bacterial species,[5,6,9] and the first detectable products are *cis*-dihydrodiols. Interest in the former stems from the realization that many aromatic hydrocarbons are carcinogenic and/or mutagenic after activation by host detoxification of these compounds, and this has received much public attention.

Several aspects of the subject of this chapter will be dealt with elsewhere in this volume, in particular the enzymology of the oxygenases used for hydroxylation and ring cleavage (Chapter 1) and the genetics and evolution of degradative functions (Chapters 5 and 6). In this chapter we will try to reduce the large number of reports on the microbial transformations of aromatic hydrocarbons into generalized catabolic pathways, realizing of course the dangers of this approach. It will become clear that evidence for most of the pathways illustrated comes from many isolated observations with different bacterial strains. Even within the same species and subtype different metabolic options are available. The exceptions to the generalized pathways that frequently occur surely emphasize the biochemical and genetic plasticity of microorganisms. We shall discuss only the pathways elaborated in microorganisms able to grow with the aromatic hydrocarbons as carbon and energy source, and this reduces the biological scope of our account to bacteria. These will include the major catabolic routes to amphibolic intermediates and some minor alternative transformations that may lead to dead-end products. The widespread abilities of microorganisms to catalyze partial transformation of aromatic hydrocarbons to products unable to support growth is extensively documented in recent publications, and will not be discussed.[7,8,11]

II. UTILIZATION OF BENZENE

Although Söhngen in 1913[20] reported the microbial utilization of benzene, and Arnaudi et al.,[21] detected catechol as a metabolite of benzene oxidation and excluded

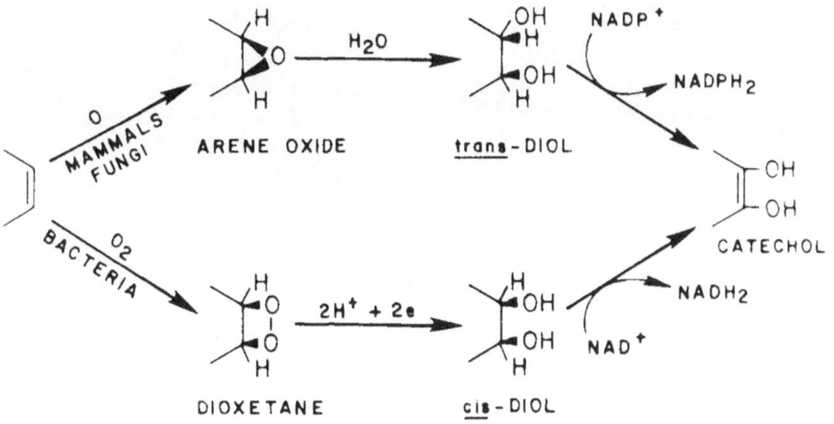

FIGURE 1. Generalized metabolic routes to catechols from aromatic hydrocarbons for eucaryotic and procaryotic organisms, modified from Gibson.[6] The dioxetane intermediate shown has not been formally demonstrated.

phenol as an intermediate, surprisingly little information about enzyme reactions unique to benzene oxidation was available prior to 1968. Other groups had noted that benzene-grown bacteria, including *Mycobacterium rhodochrous* and *Pseudomonas aeruginosa*,[25] were able to oxidize catechol. Intermediates of the *ortho*-cleavage pathway for catechol, *cis-cis*-muconate and succinate were isolated from culture media (Figure 2).[24,25]

The investigations initiated with *P. putida* by Gibson et al.[26,27] have provided definitive evidence for the formation of a *cis*-benzenedihydrodiol (*cis*-1,2-dihydro-1,2-dihydroxybenzene), and both atoms of oxygen were shown to be derived from molecular oxygen. This is in direct contrast to the fairly extensive studies of benzene metabolism in animal systems[28-30] which produce the *trans*-benzenediol by hydration of benzene epoxide (Figure 1). The evidence for the involvement of the *cis*-benzenediol in benzene metabolism by *P. putida* in the pathway shown in Figure 2 is

1. After growth with appropriate aromatic hydrocarbons, whole cell suspensions readily oxidized benzene, the *cis*-benzenediol and catechol, but not phenol nor the *trans*-benzenediol. All substrates were poorly oxidized by cells grown in the absence of hydrocarbon.
2. The stoichiometric relationships of O_2 consumption for benzene oxidation catalyzed by extracts of cells, and each of the intermediates in Figure 2 was that expected for two consecutive two-electron oxidations followed by a dioxygenative ring cleavage, i.e., 2, 1.5 and 1 mol of O_2 was consumed for each mole of benzene, *cis*-benzenediol and catechol provided, respectively.
3. Cell extracts supplemented with Fe^{2+}, cysteine, ethanol, and nicotinamide adenine dinucleotide catalyzed the oxidation of benzene, *cis*-benzenediol and catechol, and α-hydroxymuconic semialdehyde was formed in each case.
4. Isotope trapping experiments showed that unlabeled pools of *cis*-benzenediol and catechol became labeled when ^{14}C-benzene was being oxidized by cell suspensions. A pool of *trans*-diol did not become labeled.
5. A mutant of *P. putida* (strain 39/D) unable to grow with aromatic hydrocarbons as sole carbon source accumulated from benzene, a benzenediol, which was identical to a synthetic specimen of the *cis*-benzenediol. Cell extracts of the mutant strain lacked an NAD-dependent dehydrogenase for the *cis*-benzenediol which was present in the wild type.

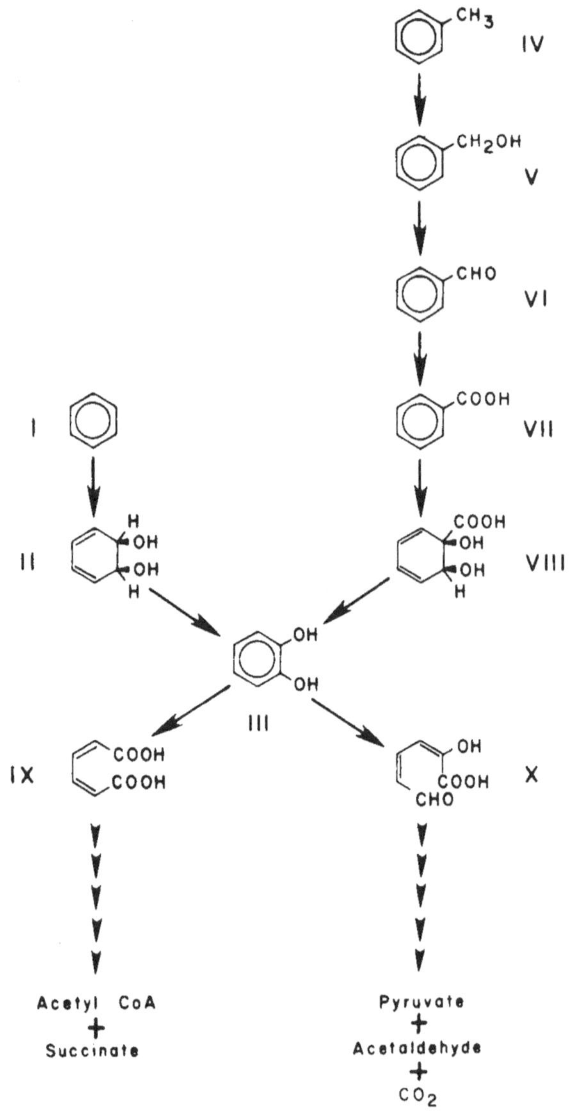

FIGURE 2. Pathways for benzene and toluene catabolism that converge and diverge with catechol as the central intermediate.

Similar lines of evidence were obtained by Högn and Jaenicke[31] to establish the same pathway from benzene to catechol in a *Moraxella* sp.; this catabolic pathway diverged from the *P. putida* pathway with an *ortho*-ring cleavage to give β-ketoadipate. Atmospheric oxygen was similarly shown to be the source of both oxygen atoms on the ring in the *Moraxella* system, but in this case catechol was the product isolated for mass analysis.

A. Enzymes of Benzene Oxidation

Enzymes for the oxidation of benzene have been extracted from a variety of bacteria including *P. putida*,[26,32] *Moraxella*,[31] and *Arthrobacter*,[33] the latter being extremely unstable. Early work showed that at least two components of the system could be separated by ammonium sulfate fractionation.[26,32] Axcell and Geary[32] obtained from

P. putida three protein fractions after chromatography of cell extracts on DEAE-cellulose. (*cis*-Benzenediol dehydrogenase and catechol 1,2-dioxygenase also present in crude extracts of this strain were separated from these three fractions and purified as by-products.) Colored proteins A1, A2, and B from each of the fractions were substantially purified to apparent homogeneity and a combination of all three, together with substrates benzene and NADH, and Fe^{2+} ions were necessary for the transformation of benzene into the *cis*-benzenediol.

Several of the properties of proteins A1, A2, and B were studied, and this allowed the formulation of an electron transport scheme between them, passing the reducing equivalents from NADH to the dioxygenase component A1 for the oxygen fixation step into benzene.

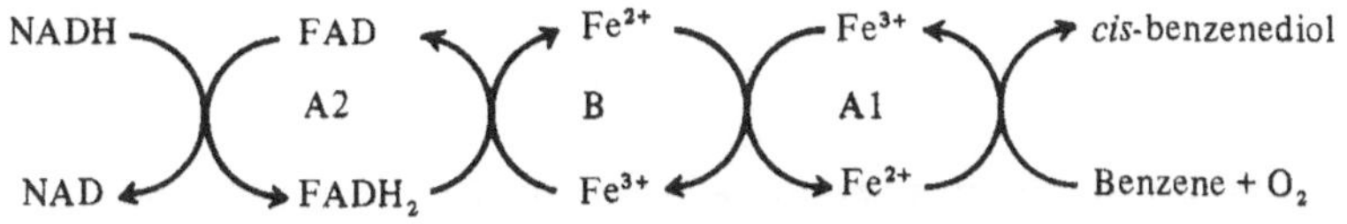

NADPH was a poor substitute of NADH in this system. Several other aromatic hydrocarbons and halo-derivatives such as toluene, fluorobenzene, chlorobenzene, bromobenzene, and iodobenzene stimulated oxygen consumption in the presence of NADH but it was not shown if they were hydroxylated.

The *cis*-benzenediol dehydrogenase had been purified previously from the same strain of *P. putida* by Axcell and Geary.[33] This enzyme was specific for both the *cis*-diol substrate and NAD. Neither *trans*-benzenediol nor NADP could participate as reactants in the catalysis. *trans*-Benzenediol and NADP were found to be the substrates for an analogous enzymic activity extracted from rabbit liver. A "*trans*-benzenediol" dehydrogenase from *Aerobacter aerogenes* was described which preferred NAD as electron acceptor for this activity of unknown function.[34] *cis*-Benzenediol dehydrogenase from *P. putida* appeared to be a tetramer of similar subunits with approximate molecular weight of 110,000. It displayed maximum catalytic activity in the presence of Fe^{2+} and glutathione, and chelating reagents, e.g., 1,10-phenanthroline or sulfydryl group reagents strongly inhibited the dehydrogenase.

III. UTILIZATION OF MONOALKYL BENZENES

Alkylbenzenes provide options for pathways of biodegradation alternative to the direct dioxygenation of the aromatic nucleus as in benzene catabolism. Thus the alkyl side chain can be oxidized prior to ring hydroxylation, or it can be retained to give homologues of the intermediates of benzene catabolism.

A. Toluene

Kitagawa[35] showed that when suspensions of *P. aeruginosa* were exposed to toluene, enzymes for the oxidation of toluene, benzyl alcohol, benzaldehyde, benzoate, and catechol were induced. This suggested that the methyl group of toluene was oxidized to benzoate prior to the formation of the ring cleavage substrate, catechol, presumably via the carboxyhydrodiol described later by Reiner and Hegeman (Figure 2).[36]

The strain of *P. putida* used by Gibson's group to study benzene catabolism described above was isolated with toluene as sole carbon source and it is also able to oxidize ethyl- and isopropylbenzene well, and propyl- and butylbenzene much less so, all giving rise to transient yellow intermediates.[26] The enrichment substrate, toluene, apparently was metabolized without the accumulation of the *meta*-ring fission product presumably because the hydrolase responsible for its dissimilation had been evolved

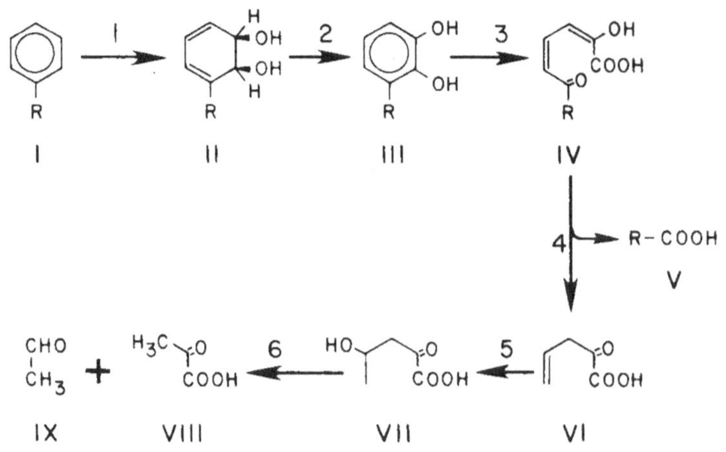

FIGURE 3. General *meta*-cleavage pathway used for the utilization of most monoalkylbenzenes (including arylbenzenes) which do not undergo oxidation at the ω-terminal carbon of the alkyl group.

for a substrate preferring a methyl group. In contrast to Kitagawa's[35] observation, *P. putida* does not oxidize benzoate; however enzymes for the oxidation of 3-methyl- and 4-methylcatechol are induced, and an initial hydroxylation of the aromatic ring seems probable. These observations are similar to those made by Claus and Walker[37] with unidentified species of *Pseudomonas* and *Achromobacter*, and those of Forro with *Nocardia corallina.*[38] Like *P. putida* these bacteria cleave catechols by *meta*-fission.

The mutant of *P. putida* 39/D described before[27] was used to accumulate a dihydrodiol from toluene; the diol was shown to be (+)-*cis*-2,3-dihydroxy-1-methylcyclohexa-4,6-diene (II, R = CH_3, Figure 3), and its absolute structure was provided by analysis of the CD spectra of the hydrogenated derivatives, *cis-cis*- and *cis-trans*-3-methylcyclohexane-1,2-diols, as 1*S*, 2*R*,[39] which was confirmed by X-ray diffraction analysis of a Diels-Alder cyclo-addition product of the diacetate of the toluene diol.[40] All the observations made are consistent with toluene being metabolized by the general *meta*-cleavage pathway shown in Figure 3.

B. Ethylbenzene

The oxidation of ethylbenzene by *P. putida* was reported by Gibson et al.[41] and a detailed study of this was made. (+)-3-Ethyl-3,5-cyclohexadiene-1-(*S*),2(*R*)-diol (II R = C_2H_5, Figure 3) was formed by the diol dehydrogenase-less strain 39/D. When this was incubated anaerobically with extracts of the wild-type strain in the presence of NAD, 3-ethylcatechol was formed. A minor diol product of ethylbenzene oxidation (about 2%), which was more polar than the diol described, was identified. This metabolite had also been oxidized at the benzylic carbon providing (+)-*cis*-3-(1′-hydroxyethyl)-3,5-cyclohexadiene-1,2-diol (III, Figure 4). There was evidence to show that this compound was formed by the stereoselective oxidation of ethylbenzene to (+)-1-phenylethanol followed by dioxygenation of the ring (Compound III, Figure 4), since the ethylbenzenediol did not serve as a precursor of this product (Compound III, Figure 4). With the wild-type strain this minor product was subject to further degradation to give *meta*-ring cleavage products which were not identified.

C. Propyl-, Propenyl-, and Butylbenzenes

A series of interesting papers has appeared over the last 6 years from Minoda's laboratory describing the bacterial utilization of propyl-, propenyl- (methylstyrenes),

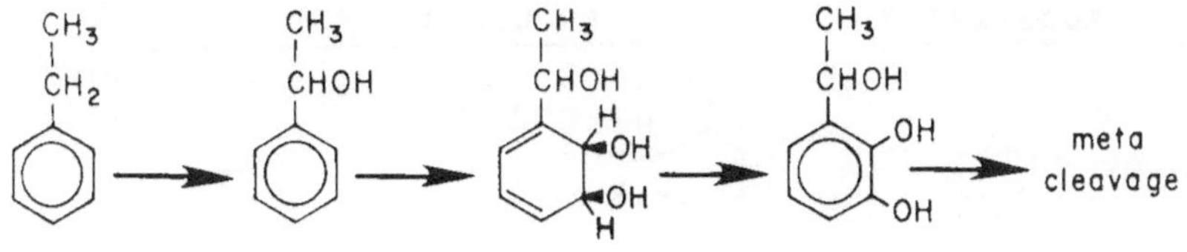

FIGURE 4. Minor metabolic transformations of ethylbenzene by *P. putida.*

and butylbenzenes. The studies are characterized by the isolation of a series of novel transformation products from culture filtrates. These products resemble intermediates of the pathway in Figure 3, but are all more reduced. They have examined two bacterial strains isolated with isopropylbenzene as carbon source, *P. desmolytica* S449B1 and *P. convexa* S107B1. These strains are able to grow with a number of *n*-alkyl, *iso*-alkyl-, *n*-alkenyl-, and *iso*-alkenylbenzenes.

After growth of *P. convexa* with α-methylstyrene, an unusual dihydrodiol was isolated and identified as (−)-*cis*-2,3-dihydroxy-1-*iso*-propenyl-6-cyclohexene (Figure 5), i.e., it is a reduced counterpart of the dihydrodiol intermediates of Figure 3.[42] This compound is formed in small quantities up to 1.5% of the α-methylstyrene supplied to a concentration of about 0.015%. The same strain after growth with α-methylstyrene yielded 3-phenylpropionaldehyde (3 g) and a little benzoate (0.5 g) from 30 g of the hydrocarbon.

P. desmolytica S339B1 was used to study *n*-butylbenzene catabolism, and an aliphatic metabolite was found to be (+)-2-hydroxy-6-oxodecanoate, i.e., analogous to an immediate ring-fission product in Figure 3, but which is saturated.[43] Similar compounds were also formed when ethylbenzene and propylbenzene were provided as carbon source.[44] When isopropylbenzene and isobutylbenzene were supplied to the bacteria, cyclic products were isolated with the same number of carbon atoms as the substrate hydrocarbon. The metabolite formed from isopropylbenzene was shown to be 2-dimethyl-1-hydroxy-3-oxo-cyclohexane-1-carboxylate and was probably formed by an intramolecular aldol condensation of a ring-fission product prior to or after a reduction of the ring (Figure 5). A similar sequence of events probably occurred in the formation of the product from *iso*-butylbenzene, with an extra event of dehydration between carbons 1 and 2 to give 2-isopropyl-3-oxo-cyclohexa-1-ene-1-carboxylate (Figure 5).[45]

The above examples did not provide evidence that side-chain oxidation is operative during growth. However the products from β-methylstyrene were phenylpropionaldehyde and benzoate, which suggested that oxidation was initiated on the *n*-alkenyl side chain, and β-oxidation gave benzoate. It is likely that all of the products which are more reduced than the intermediates of Figure 3 are end metabolic products, since their further transformation to other compounds could not be demonstrated with cell extracts or intact cells.

When *n*-propylbenzene catabolism by these strains was studied, products were obtained indicating the co-existence of transformation of the side chain prior to ring hydroxylation and vice versa.[46] Thus both strains accumulated β-phenylpropionate and benzoate indicative of β-oxidation of the side chain. They also produced 3-*n*-propylcatechol and 2-hydroxy-6-oxononanoate,[46] again a reduced counterpart of the ring-cleavage product of Figure 5. Furthermore, the former three compounds β-phenylpropionate, benzoate, and 3-*n*-propylcatechol were readily oxidized by *n*-propylbenzene grown cells. From these experiments it was concluded that both pathways play a substantial role in the utilization of *n*-propylbenzene by these strains. The effect of the

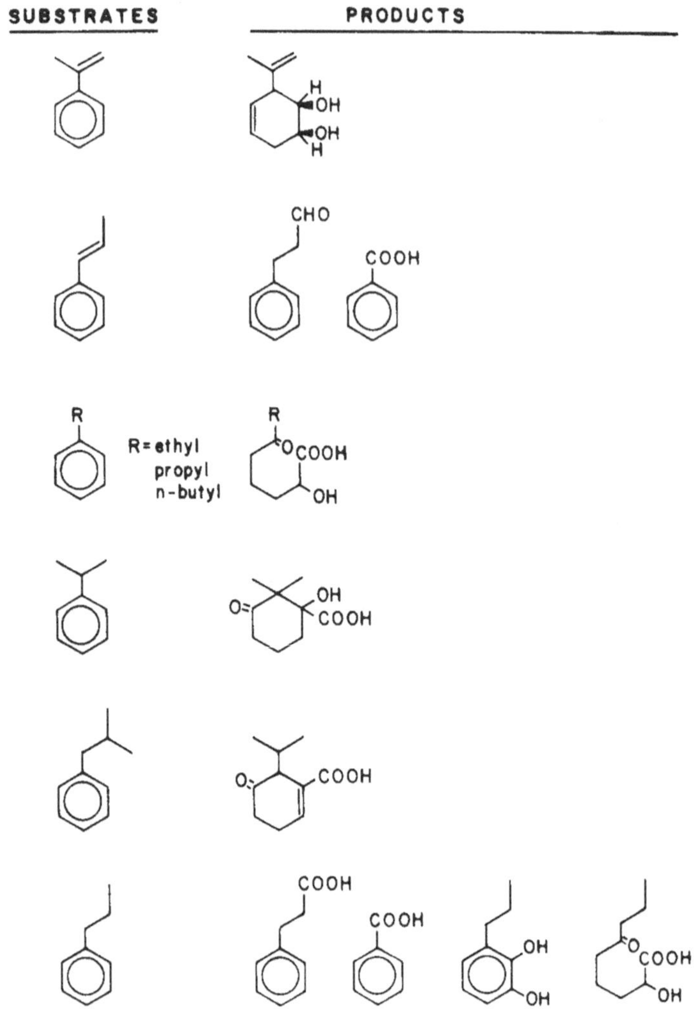

FIGURE 5. Some compounds isolated as products of metabolic transformations from various monoalkylbenzenes.

pH of the medium was tested to determine if this altered the mode of metabolism of *n*-propylbenzene (as it does for *p*-xylene by *Nocardia* sp.[47]), but no differences were detected at pH 6.0 and 7.5. The nature of the regulatory constraints on the two options is of obvious interest as one progresses from one hydrocarbon substrate to another, and no doubt this will be the subject of future research.

D. Long-Chain Phenylalkanes

The earliest studies on the utilization of long-chain phenylalkanes were made principally to determine how long-chain alkanes were metabolized,[48] i.e., the microbial counterpart of the Knoop experiments that used phenylalkanoates.[49] Long-chain phenylalkanes with an odd number of carbon atoms were oxidized by species of *Nocardia* to cinnamate and benzoate, while even-numbered carbon derivatives yielded phenylacetate.[48,50] Although both odd and even carbon chain phenylalkanes could support growth of *Nocardia* sp. the ring of even phenylalkanes was not oxidized, and presumably the release of acetyl fragments by β-oxidation supported growth. Some species of *N. salmonicolor* degraded the aromatic ring of phenylalkanes with both odd and even numbers of carbon atoms,[51] and one strain which totally oxidized both odd and even carbon phenylalkanes has been studied in detail.

N. salmonicolor (N.C.I.B. 9701) grows well with 1-phenylalkanes ranging from 1-phenylpentane to 1-phenylpentadecane (however the C-4, C-6, and C-8 alkyl groups supported poor growth).[51] Sariaslani et al. identified several products accumulated during growth with 1-phenyldodecane which indicated that β-oxidation of the side chain was not a sole and sufficient mechanism to account for the oxidation of the hydrocarbon.

Metabolites of 1-phenyldodecane oxidation amounting to 15% of the hydrocarbon supplied included: 4-phenylbutyrate, 4-phenylbut-3-enoate, 4-phenylbut-2-enoate, 3-phenylpropionate, cinnamate (4-phenylprop-2-enoate), and phenylacetate. The appearance of these metabolites was correlated with diauxic growth of *N. salmonicolor* and the specific activity of isocitrate lyase activity. These observations indicated that the alkyl side chain was preferentially metabolized by β-oxidation which supported growth, and that some alternate metabolic sequences were also used for the formation of both odd and even side-chain metabolites. The second growth phase of *N. salmonicolor* occurred after isocitrate lyase activity had decreased fivefold and phenylbutyrate and cinnamate had been exhausted.

Ortho-hydroxyphenylacetate was the only hydroxyphenylacetate formed when *N. salmonicolor* was incubated with phenylacetate, and none were found during growth with the hydrocarbon. This metabolite could also serve as a growth substrate. The pathways used for the opening of the benzene ring are not clear since 1-phenyldodecane grown cells possess elevated levels of ring-cleavage enzymes for homogentisate, 3,4-dihydroxyphenylacetate, protocatechuate, and catechol. Moreover it cannot yet be decided to what extent the α-oxidation route of phenylbutyrate via cinnamate is a salvage route since the high K_m value of phenylbutyryl CoA for a dehydrogenase is found in crude extracts of cells grown with phenyldodecane but not after growth with 4-phenylbutyrate.

When 1-phenylnonane was the growth substrate a similar diauxic growth response was observed, and metabolites that accumulated included cinnamate, phenylpropionate, benzoate, and probably *o*-hydroxyphenylpentanoate. The former two compounds were present in large amounts presumably as products of β-oxidation. Thus when odd and even phenylalkanes are metabolized there is some convergence to common intermediates, and any of several compounds could be substrates for ring cleavage. Another possible known ring-cleavage substrate for this scheme is 3,4-dihydroxycinnamate,[52] but this was not tested.

E. Biphenyl, Biphenylmethane, and Biphenylethane

Interest in the biodegradation of biophenyl and biphenylalkanes and their chlorinated derivatives (PCBs, DDT) stems from their widespread use as nuclear reactor coolants, fungistats, fire-retardants, spray stabilizers, and pesticides, and the attention they have received as environmental hazards.

Lunt and Evans[53] showed that an unidentified Gram-negative soil bacterium when growing with biphenyl as sole carbon source accumulated 2,3-dihydroxybiphenyl which was a substrate for a *meta*-cleavage enzyme. The oxidation of biphenyl and *p*-chlorobiphenyl by two strains of *Achromobacter* sp. yielded benzoate and *p*-chlorobenzoate respectively as products and led Ahmed and Focht[54] to formulate a metabolic pathway involving the *meta*-fission of 2,3-dihydroxybiphenyl (or its 4′-chloro derivative) between carbons two and three. Hydrolases of *meta*-fission pathways would give the products observed (Figure 3). Focht and Alexander[55,56] had proposed earlier a similar pathway for the utilization of biphenylmethane by *Hydrogenomonas* sp. on the basis that phenylacetate was a major degradation product and that *meta*-cleavage enzymes were constitutively expressed. Phenylacetate was metabolized later by an inducible set of enzymes; homogentisate dioxygenase was induced during growth on biphenylmethane or phenylacetate, suggesting that it was the substrate for ring fission of the phenylacetate moiety.

Formation of 2,3-dihydro-2,3-dihydroxybiphenyl from biphenyl was demonstrated with *P. putida*[57] and sufficient amounts of a diol were accumulated with a mutant of *Beijerinckia* sp. for a detailed structural study.[58] The mutant lacked an NAD-dependent dihydrodiol dehydrogenase present in the wild type that converted the *cis*-dihydrodiol of biphenyl into 2,3-dihydroxybiphenyl in the absence of oxygen. Catelani et al.,[59,60] also isolated benzoate as a metabolite of biphenyl by *P. putida* and showed that extracts of biphenyl-grown cells converted 2,3-dihydroxybiphenyl to 2-hydroxy-6-oxophenylhexa-2,4-dienoate (Compound IV, R = phenyl, Figure 3). Clearly, cleavage occurred between C1 and C2 and the data obtained with all of the bacterial strains used to date are consistent with biphenyl being metabolized via the reactions in Figure 3 involving 1,2-dioxygenative *meta*-cleavage of 2,3-dihydroxybiphenyl.[151] Further evidence that this is an important general metabolic pathway is provided by the results of Francis et al.,[61,62] who showed that a pseudomonad isolated with 1,1-diphenylethane (the DDT carbon skeleton) as sole carbon source could transform some chlorinated derivatives of the enrichment substrate to phenylpropionates, e.g., 2-(*p*-chlorophenyl)-propionate from 1-(*p*-chlorophenyl)-1-phenylethane; and 2-phenylpropionate was formed transiently from the parent hydrocarbon 1,1-biphenylethane.[62]

F. Secondary Alkylbenzenes (2,3- and 4-Phenylalkanes)

We have already discussed the catabolism of the lowest homologue of this series, i.e., isopropylbenzene in a previous section. Kallio,[63] Treccani,[64] and colleagues have provided most of the information available on this group of compounds. Studies on these and the long linear alkyl benzenes are important because of their large-scale manufacture and use of their derivatives as detergents and agricultural chemicals. McKenna and Kallio[9,63,65] showed that the position of the phenyl group on the alkane chain alters the biodegradability of this series of hydrocarbons. When the phenyl group was furthest away from a terminal carbon then the least amount of biodegradation occurred.

Baggi et al.,[64] studied the catabolism of 2-phenylbutane, 3-phenylpentane, 3-phenyldodecane, and 4-phenylheptane by *P. acidovorans*, isolated with 3-phenylpentane, *P. putida* isolated with biphenyl, and two strains of *Nocardia* sp. isolated with 3-phenyldodecane. The two pseudomonads could grow with the three shorter chain phenylalkanes and all gave rise to intermediates of Figure 3. Dihydrodiols were identified from 2-phenylbutane and 3-phenylpentane, and detected from 4-phenylheptane. A *meta*-ring-fission product, 2-hydroxy-7-ethyl-6-oxonon-2,4-dienoate (Compound IV, R = 2-ethylpropyl, Figure 3) was isolated from 3-phenylpentane. Thus the position of ring cleavage is also the same as that shown in Figure 3.

The *Nocardia* strains were not capable of growth with the above three compounds. All the evidence indicated that 3-phenyldodecane was metabolized by side-chain oxidation to four homologous acids, 2-phenylbutyrate, 3-phenylpentanoate, 4-phenylhexanoate, and 5-phenylheptanoate. This is an unusual spectrum of metabolites and uncharacteristic of expected intermediates of β-oxidation. What is clear though is that the *Nocardia* can initiate the side-chain oxidation from the longer alkyl chain only. Respiratory measurements with resting cells of *Nocardia* showed that none of the phenylalkanoates isolated above nor 3-phenyldodecane and benzoate stimulated respiration.

G. *Tert*-Butylbenzene

Compounds with quaternary carbons such as *tert*-butylbenzene are of considerable interest because they rarely occur as natural products and they are synthesized by industry for agricultural (pesticides) and domestic (detergents, surfactants) uses. Furthermore McKenna and Kallio[9,65] have described such molecules as strongly resistant to biodegradation. During a study of the metabolism of quaternary carbon compounds Catelani et al.[66] isolated a strain of *Achromobacter* that grew well with *tert*-butylbenzene. They showed that a dihydrodiol which gave 2- and 3-*tert*-butylphenol after heating in acid transiently accumulated in culture media. 2,3-Dihydroxy-*tert*-butylbenzene was a substrate for ring cleavage giving a product that absorbed at 410 nm and reacted with ammonium acetate to give a new λ max at 275 nm, consistent with the formation of a picolinate. It was not clear how readily the ring-fission product was metabolized by the hydrolase presumed to be responsible for this task; this is of interest because the *tert*-butyl ketone is a very poor substrate for hydrolases in other *meta*-cleavage pathways.[67] However, it was clear that the expected product, pivalate (2,2,2-trimethylethanoate), was formed in cultures. No evidence was obtained for the utilization of pivalate by growing or washed cells.

IV. UTILIZATION OF DIALKYLBENZENES

m- and *p*-Xylenes (and Toluene)

The methyl groups of *m*-xylene, *p*-xylene, 1,2,4-trimethylbenzene, and toluene were shown to be oxidized to carboxyls by a variety of bacteria, principally pseudomonads, isolated for their ability to grow with *m*- or *p*-xylene.[68-70] Thus *m*-xylene, *p*-xylene, 1,2,4-trimethylbenzene, and toluene gave *m*-toluate,[68,69] *p*-toluate,[69,70] 3,4-dimethylbenzoate,[69] and benzoate,[35] respectively. In addition a variety of phenolic compounds were extracted from culture filtrates, including 3-methylsalicylate from *m*-xylene,[69], *p*-cresol from *p*-xylene,[71] and 4-methylcatechol from *p*-xylene.[72] The former two products possibly arose as artifacts of diol dehydrations (see later). Davis et al.[68] showed that *meta*-ring cleavage enzymes were induced after growth on either of the xylenes, and suggested that 4-methylcatechol was formed as the ring-cleavage substrate for *p*-xylene oxidation, but they were not able to determine which of 3- and 4-methylcatechol was the ring-cleavage substrate for *m*-xylene degradation.

The discovery of the coexistence of the hydrolase[73] and the oxidative[74,75] routes for the further metabolism of the *meta*-cleavage product of catechol[75-78] (Figure 6) led Williams' group to study the metabolism of benzoate and *m*- and *p*-toluate in *P. arvilla* strain mt-2, now known as *P. putida* biotype A, PaWI; this strain was of some interest because unlike other strains in this subclass, it possessed the *meta*-cleavage enzymes for the oxidation of catechol during growth with benzoate.[79] Both branches of the *meta*-cleavage pathway were demonstrated after growth of *P. putida* PaWI with benzoate or the toluates.[79] It was also shown that 3-methylcatechol was the ring-cleavage substrate during *m*-toluate oxidation by accumulating the ring-cleavage product 2-hydroxy-6-oxohepta-2,4-dienoate (Compound VII, $R = H$, $R^1 = CH_3$, Figure 6) from *m*-toluate with washed cell suspensions at pH 5.5.[79]

A pseudomonad, *Pseudomonas* Pxy, isolated on *p*-xylene, was able to grow on toluene, *p*-methylbenzyl alcohol, *p*-methylbenzaldehyde, and *p*-toluate as well as *m*-xylene and the equivalent *m*-methyl oxidation products.[80] Two mutant organisms, unable to grow on these compounds, were isolated by Davey and Gibson. Mutant Pxy-40 grown with succinate in the presence of *p*-xylene accumulated *p*-toluate in the medium. Mutant Pxy-82 accumulated catechols from the hydrocarbons: 4-methylcatechol was formed from *p*-toluate and *p*-xylene, and 3-methylcatechol (and 3-methylsalicylate) from *m*-xylene. Figure 6 shows a generalized version of the pathways used for the metabolism of several hydrocarbons and alkylbenzoates. The top part of the pathways (to Compounds VI) were established with the blocked mutants of *P.* Pxy and the bot-

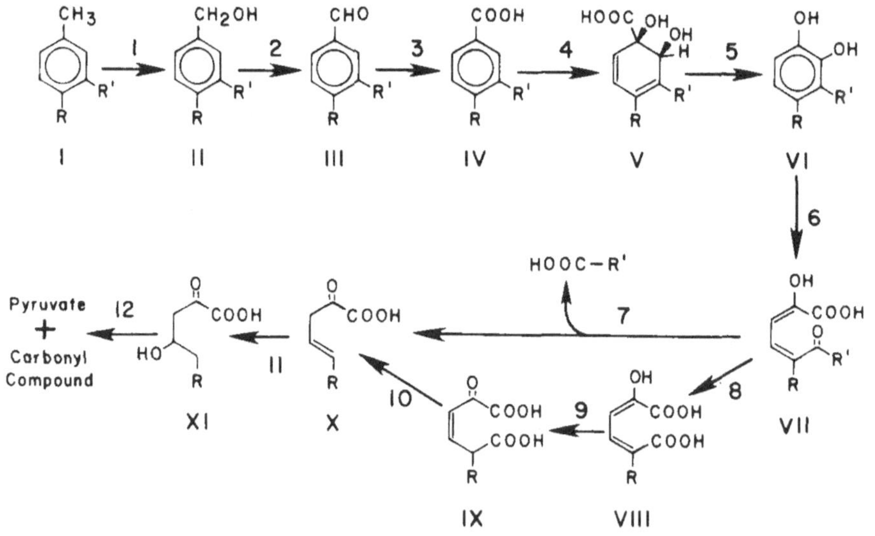

FIGURE 6. General catabolic routes for the utilization of *m*- and *p*-xylenes [and other identified TOL (XYL) functions]. R = H, R' = H; R = H, R' = CH_3; R = H, R' = C_2H_5; R = CH_3, R' = H; R = CH_3, R' = CH_3.

tom half (from Compounds IV) with the *P. putida* PaWI strain grown with the benzoates. The formation of the carboxyhydrodiols as intermediates between the benzoates and catechols is inferred from Reiner and Hegeman's studies on benzoate and several of its substituted derivatives by *Alcaligenes eutrophus*,[82] and the isolation of 3-methylsalicylate from *m*-xylene,[69,80] probably due to an acid catalyzed dehydration of the carboxyhydrodiol during extraction of the intermediates.

Another very interesting aspect about the formulation of this pathway was the independent discovery by three groups that the ability to grow with the toluates was a function of the presence of a transmissible plasmid, TOL, in *P. arvilla* mt-2 (*P. putida* PaWI).[83-85] Worsey and Williams then demonstrated that *P. arvilla* mt-2, and later many other strains isolated with *m*-toluate, were able to grow with toluene, and *m*- and *p*-xylenes,[86,87] and that all functions of the pathways in Figure 6 were plasmid encoded (TOL), as were those in *P.* Pxy[88] (XYL). A discussion of these plasmids is outside the scope of this chapter, but is presented elsewhere in this volume.

The legend of Figure 6 shows that the TOL (XYL) functions are much less specific than formerly demonstrated. Kunz and Chapman have shown in *P. arvilla* mt-2 and two strains of *P. putida* that TOL pathway functions are also used for growth at the expense of 1,2,4-trimethylbenzene, 3-ethyltoluene, 3,4-dimethylbenzoate, 3-ethylbenzoate,[89,90] 4-ethyl-3-methyltoluene, and 4-ethyl-3-methylbenzoate,[91] and some of the alcohol and aldehyde intermediates.

B. 4-Alkyltoluenes

Aside from the studies of *p*-xylene (4-methyltoluene) metabolism, which have already been discussed, *p*-cymene (4-isopropyltoluene) seems to be the only other compound of this 1,4-dialkyl-substitution pattern which has been examined. This possibly is related to the origin and importance of the cymene carbon skeleton in a vast array of natural products including many fragrances and flavors. The extensive studies of Bhattacharyya and colleagues on the microbiological degradation of a series of alicyclic terpenes such as the pinenes, limonene, and menthene derivatives led to the observation that *p*-cymene could also serve as carbon and energy source for *Pseudomonas*

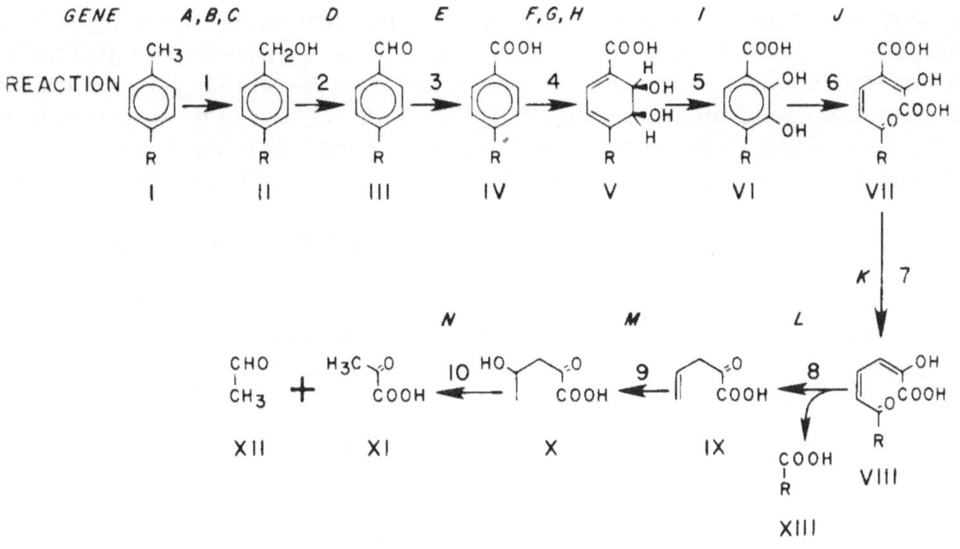

FIGURE 7. General pathway for the catabolism of *p*-alkyltoluenes (including aryltoluenes). R = methyl, ethyl, *n*-propyl, *iso*-propyl, *n*-butyl, *tert*-butyl, pentyl, phenyl, benzyl.

sp. PL[92-94] (later identified as *P. putida*),[97,98] a strain isolated with limonene as carbon source. *P. putida* PL was shown to initiate catabolism of *p*-cymene by oxidation of the ring methyl substituent to carboxyl to give *p*-cumate (4-isopropylbenzoate) as had been observed earlier by Omori et al.[95] and by Leavitt[96] for other pseudomonads. 2,3-Dihydroxy-*p*-cumate was identified as a substrate for ring cleavage,[93,94] i.e., the carboxyl group was retained for ring cleavage. DeFrank and Ribbons confirmed these observations with *P. putida* PL, and with the aid of two mutants defective in *p*-cymene catabolism documented the remainder of the pathway shown in Figure 7. One of the mutants accumulated 2,3-dihydro-2,3-dihydroxy-4-alkylbenzoates from a variety of 4-alkyltoluenes or benzoates including *p*-xylene and *p*-toluate.[97,98] The other mutant was particularly useful because it was defective in a decarboxylase responsible for transforming the immediate ring-fission product into the substrate for a hydrolase (Reaction 8, Figure 7).[99] The *p*-cymene pathway provided the second example of 2,3-dihydroxybenzoate(s) being used as *meta*-ring-cleavage substrates[100] and confirmed that the position of ring cleavage was between carbons 3 and 4 of the ring (Figure 7).[99] The specificity of enzymes of the pathway, like those for the TOL systems, are very relaxed, as are the regulatory constraints imposed on them.[98,99]

Wigmore et al.[101] examined the possibility that 2,3-dihydroxy-4-alkylbenzoates might be a general class of ring-cleavage substrates for the utilization of other 4-alkyltoluenes and benzoates. Several species of *Pseudomonas* were isolated from enrichments made with various 4-alkyltoluenes or benzoates as sole carbon source. Isolates included strains of *P. fluorescens, P. putida, P. aeruginosa* and *P. stutzeri.* All of 30 cultures isolated with 4-ethyl, 4-propyl-, 4-isopropyl-, 4-butyl-, 4-pentyl-, 4-phenyl-, and 4-benzyltoluenes or benzoates readily oxidized 2,3-dihydroxy-*p*-cumate, but not 4-alkylcatechols after growth on their enrichment substrates. By contrast, species isolated with *p*-xylene, *p*-toluate, 1,2,4-trimethylbenzene or 3,4-dimethylbenzoate did not oxidize 2,3-dihydroxy-*p*-toluate but readily oxidized 3- and 4-methylcatechols. The conclusions reached from this survey of respiratory patterns were (1) enrichments made with *p*-xylene or *p*-toluate yielded strains that oxidized these compounds via 4-methylcatechol as ring-cleavage substrate (Figure 6), and (2) enrichments made with 4-alkyl toluenes and -benzoates with the 4-alkyl group larger than methyl invariably yielded strains that used 2,3-dihydroxy-4-alkylbenzoates as ring-cleavage substrates (Figure 7). A detailed biochemical and genetic analysis showed that the pathway in Figure 7 exists in several strains of the latter group.[102] This pathway appears to be another general *meta*-fission route, differing from those in Figures 3 and 6 by using (1) 2,3-dihydroxybenzoates as ring-fission substrates and (2) a decarboxylase (Reaction 7, Figure 7).

V. UTILIZATION OF FUSED POLYNUCLEAR HYDROCARBONS

The three simplest fused polynuclear hydrocarbons, naphthalene, anthracene, and phenanthrene (and some of their methyl analogs), are the only compounds of this class that have been shown to support the growth of microorganisms. Initial interest in their metabolism was centered on the processes used for detoxication of foreign compounds in the diets of laboratory rodents.[28,103] Booth and Boyland[104] established that *trans*-dihydrodiols of polynuclear hydrocarbons were metabolites, and this possibly influenced for many years the assignment by analogy of the stereochemistry of dihydrodiols isolated as bacterial metabolites. Several of the more complex fused polynuclear hydrocarbons, e.g., the benzanthracenes, which are known to be pro-carcinogens are substrates for microbial hydroxylations, both procaryotic and eucaryotic, but none of them have been shown to support growth when presented as sole carbon source. This topic has been recently reviewed.[105]

A. Naphthalene

Studies on the decomposition of naphthalene in soils date back to the 1920s when Tattersfield[106] observed that microbial activities were at least in part responsible. Gray and Thornton[107] and Tausson[108] isolated several bacterial strains able to utilize naphthalene, but intermediary metabolites for naphthalene dissimilation were not elucidated until 25 years had elapsed. Then three groups showed that salicylate was formed in cultures of various bacteria isolated by enrichment with naphthalene. Walker and Wiltshire[109] established that salicylate was formed as a product of naphthalene oxidation and that it was utilized by naphthalene grown cells of a soil bacillus. Salicylate had previously been reported as a major product of naphthalene oxidation with a pseudomonad isolated by Strawinski and Stone.[110] A 1,2-dihydro-1,2-dihydroxynaphthalene was also isolated from culture filtrates of the soil bacillus, and a pathway for naphthalene oxidation through this compound to salicylate and then catechol was formulated for the bacillus, a *Nocardia* strain, and three pseudomonads.[111,112] It was suggested that the dihydrodiol which was isolated in only very small quantities had the D-*trans*-stereochemistry on the basis of similar optical rotation to that of the D-*trans*-isomer isolated as a metabolite from rat urine. Subsequent reports[113-117] made similar suggestions based on reference to the original assignment.

The naphthalene dihydrodiol accumulated from naphthalene by a mutant of *Pseudomonas* sp. was rigorously identified as (+)-*cis*-1(*R*),2(*S*)-dihydroxy-1,2-dihydronaphthalene by Gibson, Jerina et al. (Compound II, Figure 8).[118,119] This *cis*-dihydrodiol and a *trans*-isomer co-chromatographed in several solvent systems but were easily distinguished by their NMR spectra and dehydration rates in acid to give mainly 1-naphthol, the *cis*-diol being about 50 times less stable than the *trans*-isomer.[119] The *cis*-diol was shown by William's group to be the preferred isomer for oxidation by whole cells and extracts of *Pseudomonas* sp. (N.C.I.B. 9816), a strain used earlier.[120]

Several aspects of the initial reactions of naphthalene oxidation by pseudomonads have been explored.[119] Five strains, including three formerly studied, were shown to form the *cis*-diol by isotope trapping experiments. When ^{14}C-naphthalene oxidation

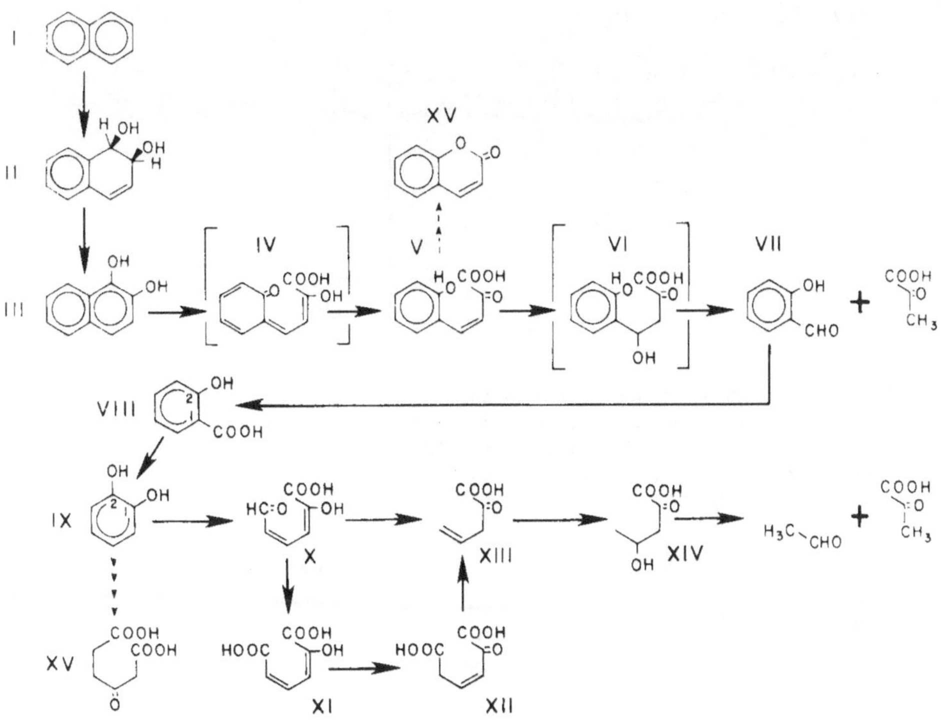

FIGURE 8. Pathways for naphthalene catabolism which diverge at catechol (IX), and α-hydroxymuconic semialdehyde (X).

was catalyzed by extracts of all the pseudomonads in the presence of pools of carrier *cis-* and *trans-*diols, significant radioactivity was detected only in the *cis-*isomer; the *trans-*diol became labeled when rat liver microsomes replaced the bacterial extracts. The initial hydroxylation of naphthalene to give the *cis-*diol in *P. putida* was dependent on NADH or NADPH as reductant. When NADPH was provided the *cis-*diol accumulated and was only further transformed if NAD was provided. The latter reaction was not catalyzed by extracts of the mutant that accumulated the *cis-*diol. The *cis-*diol was formed by dioxygenation since the parent ion composition of the isotopic species in the product (M, M + 2, and M + 4) was the same as in the oxygen (O_2) used for the incubation.

Fernley and Evans[114] isolated coumarin (XV, Figure 8) from culture filtrates of a pseudomonad growing with naphthalene. This compound was detected by its characteristic odor, which was largely responsible for its rapid identification. Coumarin was also extracted as a product of 1,2-dihydroxynaphthalene oxidation catalyzed by extracts of the cells. This led to the proposal that 1,2-dihydroxynaphthalene was the substrate for a ring-cleavage enzyme to give, putatively, *o-*carboxy-*cis-*cinnamate by fission between C1 and C2. This could give rise to *o-*hydroxy-*cis-*cinnamate by oxidative decarboxylation. The latter compound readily formed coumarin on acidification and could conceivably be the precursor of salicylate by reactions of β-oxidation.

Using a different pseudomonad Davies and Evans[121] demonstrated that 1,2-dihydroxynaphthalene was oxidized by *meta-*ring fission, between C1 and the angular C to give *cis-o–*hydroxybenzalpyruvate (V, Figure 8) which can exist in several forms, some of which are shown in Figure 9. Both *cis-* and *trans-o–*hydroxybenzalpyruvate were converted to salicylaldehyde by cell-free extracts, and an NAD-dependent dehydrogenase catalyzed the formation of salicylate. Further metabolism of salicylate was

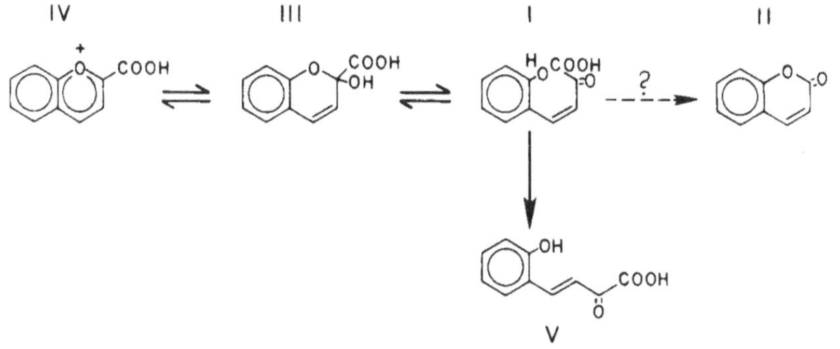

FIGURE 9. Some of the interconvertible chemical species of 2-hydroxybenzalpyruvate (I - *cis*-; V-*trans*). II, coumarin; III, hemiacetal form of 2-hydroxybenzalpyruvate (2-carboxy-2-hydroxychromene); IV, cationic form of III. Interconversion of the *cis* and *trans* isomer of 2-hydroxybenzalpyruvate is presumed to occur via the "quininoid" shown as Compound IV in Figure 8.

presumed to proceed via catechol since α-hydroxymuconic semialdehyde was shown to be a product.

Figure 8 is a composite of the observations made for naphthalene degradation by many different bacterial strains. Both the hydrolytic and oxidative routes for the utilization of α-hydroxymuconic semialdehyde (X, Figure 8) function in one pseudomonad, at least after growth on naphthalene.[122-124] The figure has been constructed to show possible mechanistic similarities used for the oxidation of each of the rings. It had been suggested that some of the enzymes were nonspecific so that they catalyzed analogous reactions such as ring cleavage (III → IV and IX → X, Figure 8), hydration (V → VI and XIII → XIV, Figure 8), and aldol fission of VI and XIV (Figure 8) and X (Figure 7). There is no evidence for this proposal and separate enzymes for the cleavage of 1,2-dihydroxynaphthalene and catechol have been found in some strains grown with naphthalene.[125,126] Furthermore, both unilateral expression of *ortho*- and *meta*-cleavage pathways for catechol and their co-existence appear to be catabolic options that are strain dependent (see later).[124,125,127-131]

B. Methylnaphthalenes

The known intermediates of 1- and 2-methylnaphthalene catabolism, identified for organisms that grow with these compounds, are summarized in Figure 10. Treccani and Fiecchi[132,133] first reported that a methylnaphthalene dihydrodiol and a methylsalicylate were products of 2-methylnaphthalene catabolism by *P. desmolyticum*, a strain isolated with naphthalene. The dihydrodiol was identified as 1,2-dihydro-1,2-dihydroxy-7-methylnaphthalene which was the precursor of 4-methylsalicylate.[134] Other metabolites which accumulated during growth of *P. desmoliticum* with 2-methylnaphthalene included 2-hydroxymethylnaphthalene (V, Figure 10), 2-naphthoate, and 4-hydroxymethylsalicylate (VII, Figure 10). The two latter compounds are considered to be dead-end products of 2-methylnaphthalene catabolism. Rogoff and Wender[135] similarly showed that 4-methylsalicylate was formed from 2-methylnaphthalene by *Pseudomonas* sp. PM-1, a strain isolated for phenanthrene utilization, as well as *Pseudomonas* sp. N-1-a, which was obtained by enrichment with naphthalene. Thus with the organisms studied so far, and isolated on naphthalene or phenanthrene, the two methyl naphthalenes are metabolized by dioxygenation of the unsubstituted ring followed presumably by a reaction sequence analogous to that for naphthalene catabolism. As with strains that have been grown with naphthalene both options of ring

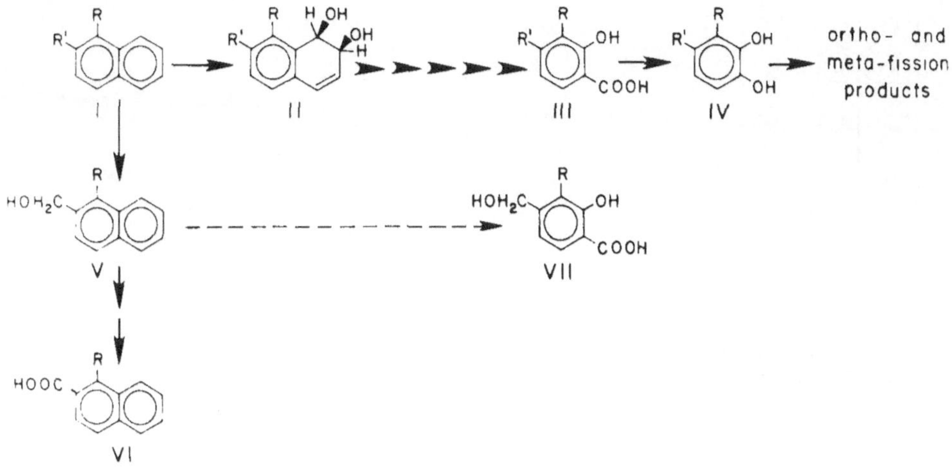

FIGURE 10. Proposed pathways for 1- and 2-methylnaphthalene utilization. 1-Methylnaphthalene: R = CH_3, R^1 = H; 2-methylnaphthalene: R = H, R^1 = CH_3.

cleavage occur for catechol and methylcatechols that are formed, and this depends on the strain under study. Thus naphthalene-grown *P. desmolyticum* possessed both *ortho-* and *meta-*fission pathways for catechol(s) cleavage.[124] On the other hand, Azoulay showed that only the *meta-*cleavage pathway was present in the naphthalene-grown pseudomonad he used, although the *ortho-*cleavage pathway for catechol was used after growth with benzoate.[128] The surveys that Barnsley has made on the regulation of enzyme synthesis summarize the data on other strains[126,129-131] (see also Reference 127). An observation concerning ring cleavage by naphthalene-grown *P. desmolyticum* was that the *meta-*cleavage enzyme catechol 2,3-dioxygenase had very low activity toward catechol so that catechol was channeled down the *ortho-*pathway in this strain. Another pathway for 4-methylcatechol oxidation by *P. desmolyticum* may also be available. After maintenance of this strain on 5-methylsalicylate the ability to synthesize catechol 2,3-dioxygenase was lost.[136] Such cells were then able to metabolize 4-methylcatechol to β-ketoadipate, and protocatechuate was shown to be an intermediate. Thus the utilization of 2-methylnaphthalene may occur via oxidation of the methyl group of 4-methylcatechol to protocatechuate followed by *ortho* fission, since protocatechuate 3,4-dioxygenase was induced after growth with naphthalene.

C. Phenanthrene

Figure 11 is another example of a composite scheme to accommodate several independent and specific observations made with different bacteria able to grow with phenanthrene. Metabolic studies were initiated by Colla et al.[137,139] and Rogoff and Wender,[138,140] who showed that 1-hydroxy-2-naphthoate (VIII, Figure 11) was a product of phenanthrene oxidation by a *Flavobacterium* and *P. aeruginosa*, respectively. Colla et al. also isolated a 3,4-dihydro-3,4-dihydroxyphenanthrene and salicylate from culture filtrates of the *Flavobacterium*. Both groups showed that phenanthrene grown cells rapidly oxidized salicylate and catechol; 1-hydroxy-2-naphthoate was oxidized slowly with carbon dioxide evolution. Mutants of *Beijerinckia* sp. B836 and *P. putida* 119, that had been characterized for their ability to accumulate *cis-*dihydrodiols from biphenyl[58] and naphthalene,[118] respectively, were able to oxidize phenanthrene to (+)-*cis-*3,4-dihydroxy-3,4-dihydrophenanthrene (II, Figure 11) as the principal product.[141] The configuration (3*S*,4*R*) assigned to this metabolite was the first determination of the absolute stereochemistry of a bay region dihydrodiol.[142] A minor dihydrodiol me-

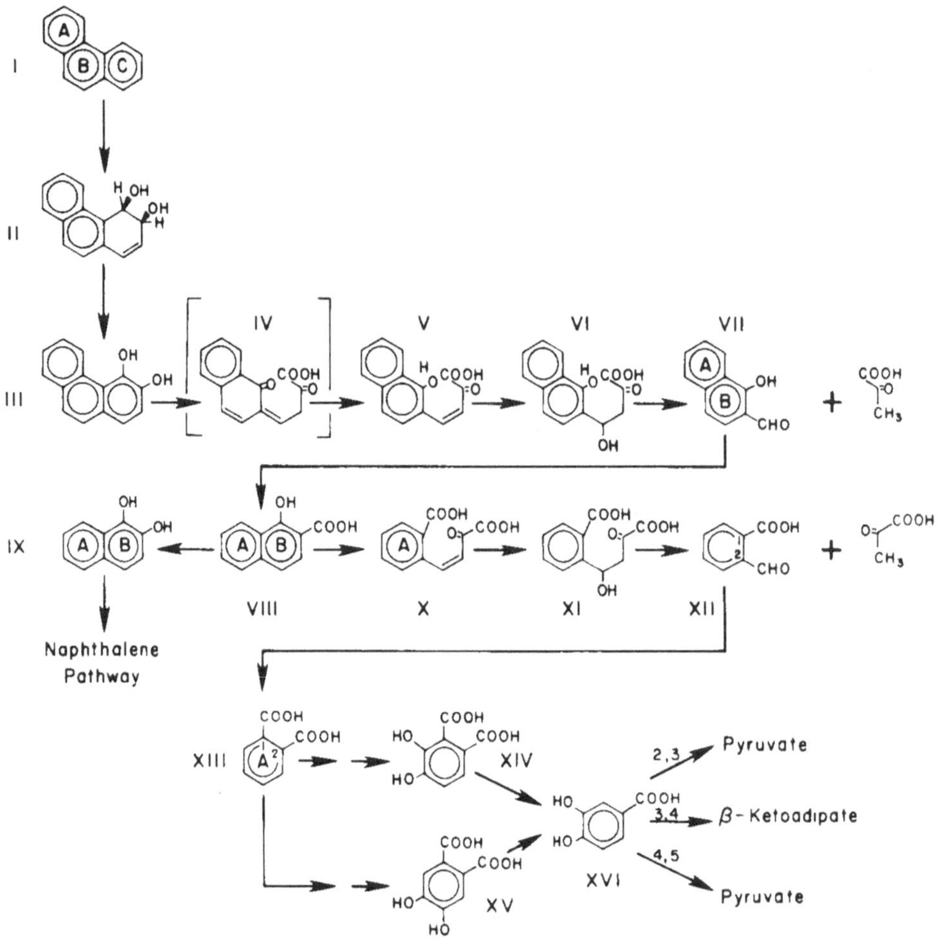

FIGURE 11. A network of pathways available for phenanthrene catabolism. Potential branch point metabolites are 1-hydroxy-2-naphthoate (VIII), *o*-phthalate (XIII), protocatechuate (XVI), as well as catechol and hydroxymuconic semialdehyde which are not shown; see naphthalene branch pathway detailed in Figure 8.

tabolite (about 5 to 10%) was shown to be *cis*-1,2-dihydro-1,2-dihydroxyphenanthrene with the (1*R*,2*S*) configuration.[142]

Several of the intermediary metabolites of phenanthrene catabolism between *cis*-3,4-dihydro-3,4-dihydroxyphenanthrene and 1-hydroxy-2-naphthoate were elucidated by Evans et al.[143] with three pseudomonads, *P. aeruginosa*,[139] and two unidentified strains $P._F$ and $P._G$ isolated by enrichment with naphthalene as carbon source.[121] 3,4-Dihydroxyphenanthrene was rapidly oxidized by cells grown with phenanthrene or naphthalene with the transient accumulation of yellow intermediate(s). In addition to compounds previously isolated as metabolites, or shown to be oxidized, 1-hydroxy-2-naphthaldehyde and 1,2-dihydroxynaphthalene were readily oxidized by phenanthrene- (or naphthalene-?) grown cells. An NAD-dependent dehydrogenase was shown to convert 1-hydroxy-2-naphthaldehyde to 1-hydroxy-2-naphthoate, and it was postulated that the latter was the precursor of 1,2-dihydroxynaphthalene (Figure 11).

The similarity of the proposed reaction sequences leading from phenanthrene to 1,2-dihydroxynaphthalene (Figure 11) and naphthalene to catechol (Figure 8) led Ribbons[144] to propose that they may be catalyzed by a common suite of nonspecific enzymes. There is little evidence to support this, and the flux of phenanthrene carbon atoms through various metabolite pools is probably far more complex than can be accounted for by a single pathway, as discussed later.

Kiyohara et al.[145-147] have studied phenanthrene and naphthalene catabolism in a variety of bacteria including *Aeromonas* sp., fluorescent and nonfluorescent pseudomonads, *Vibrio* sp., and some unidentified species. The initial observations with a nutritionally versatile *Aeromonas* sp. S45P1 suggested that *o*-phthalate and protocatechuate were intermediates (Figure 11), and not salicylate and catechol as proposed previously. A ring-cleavage enzyme for 1-hydroxy-2-naphthoate was found in extracts of phenanthrene-grown cells and the product absorbing at 300 nm was further transformed to phthalaldehydate[146] (XII, Figure 11). The latter was converted to phthalate in the presence of NAD. Thus cleavage of 1-hydroxy-2-naphthoate between C1 and C2 had occurred; subsequent hydration and aldol fission would account for the formation of phthalaldehydate. The ring cleavage is novel because it is one of the rare examples of the substrate possessing only one hydroxyl group. 1,4-Dihydroxy-2-naphthoate, a substrate for gentisate 1,2-dioxygenase, was a poor substrate for enzymes in phenanthrene-grown cells.[146] Another interesting feature of the enzyme is that it may cleave 1,2-dihydroxynaphthalene between C1 and the angular C, since *cis*-benzalpyruvate is apparently formed from naphthalene by phenanthrene-grown cells. It has been shown that catechol 1,2-dioxygenase from *Brevibacterium* sp. will cleave 3-methylcatechol between C1 and C2, and C2 and C3.[74]

Fourteen bacterial strains independently isolated by enrichment with phenanthrene, grew with *o*-phthalate, but only three of these grew with naphthalene.[147] In all of the phenanthrene-grown cells 1-hydroxy-2-naphthoate, 2-formylbenzoate, *o*-phthalate, and protocatechuate stimulated respiration better than salicylaldehyde, salicylate, or catechol.[147] However, levels of salicylate hydroxylase, catechol-1,2-dioxygenase and catechol 2,3-dioxygenase were insignificant in phenanthrene-grown cells. After growth of three of the strains with naphthalene, salicylate and catechol were rapidly oxidized, whereas 1-hydroxy-2-naphthoate, phthalaldehydate, *o*-phthalate, and protocatechuate were poor substrates.

Thus there appear to be several metabolic routes available for phenanthrene catabolism. This is evident from the following:

1. Sequential induction studies had shown that different enzyme complements are expressed in *P. aeruginosa* after growth with phenanthrene or naphthalene, and overlapping substrate specificities were evident for several polycyclic hydrocarbons.
2. Two *cis*-dihydrodiols (1,2- and 3,4) are formed as the immediate oxidation products of phenanthrene.
3. Two bicyclic compounds (1,2-dihydroxynaphthalene and 1-hydroxy-2-naphthoate) are substrates for ring cleavage in phenanthrene-grown cells.
4. Catechol 1,2-dioxygenase and catechol 2,3-dioxygenases are present in some strains.
5. The oxidative and hydrolytic routes co-exist after 2,3-dioxygenative cleavage of catechol.
6. Protocatechuate 3,4-dioxygenase and protocatechuate-4,5-dioxygenase are present in cell extracts in several strains after growth with phenanthrene.[147] All of these observations suggest that phenanthrene catabolism in some strains may be very complex and diversified due to the low specificities of the enzymes expressed, as well as the regulatory responses elicited, such that several metabolic pathways provide a network of paths for degradation. Individual strains and environmental or cultural conditions undoubtedly add to the variety of options available.

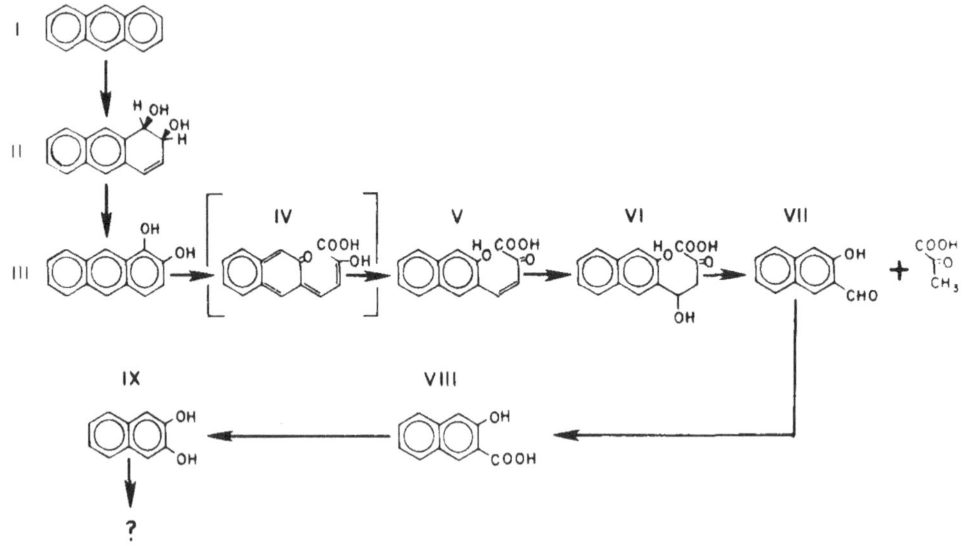

FIGURE 12. Proposed catabolic pathway for anthracene catabolism.

D. Anthracene

Little is known about the degradation of anthracene. Like the studies made with phenanthrene interpretations of observations made are difficult because of the uncertainties of which pathways are operating. Figure 12 is the usual catabolic pathway described for anthracene, but it should be regarded with caution.

3-Hydroxy-2-naphthoate was isolated as a metabolite of anthracene with cultures of *P. aeruginosa*[148] and *Flavobacterium.*[137] Anthracene grown cells of *P. aeruginosa* and $P._{G}$[143] readily oxidized salicylate and catechol, but it is not known if this was because of the nonspecific induction of a naphthalene pathway by anthracene. 1,2-Dihydroxyanthracene and 3-hydroxy-2-naphthaldehyde were readily oxidized by anthracene- or naphthalene-grown cells of $P._{G}$. Because of this naphthalene was also used as the growth substrate to give good cell yields, required for metabolic studies. The cross-specifities of induction by, and oxidation of polynuclear hydrocarbons[143] provided the basis of this approach.

(+)*cis*-1,2-Dihydro-1,2-dihydroxyanthracene was isolated from culture filtrate of the diol dehydrogenase-less mutant B836 of *Beijerinckii* sp. The absolute configuration was shown to be (1 *R*,2 *S*).[150]

Extracts from naphthalene-grown $P._{G}$ readily oxidized 1,2-dihydroxyanthracene to 2-hydroxy-3-naphthoate. When extracts were diluted, a compound with properties consistent with *cis*-4-(2-hydroxynaphth-3-yl)-2-oxobut-3-enoate (Compound V, Figure 12) accumulated, which like its analog produced from naphthalene, could be converted to a coumarin derivative. An NAD-dependent enzyme(s) oxidized 3-hydroxy-2-naphthaldehyde as well as 1-hydroxy-2-naphthaldehyde and salicylaldehyde, the equivalent metabolites of phenanthrene and naphthalene degradation, respectively. 3-Hydroxy-2-naphthoate was proposed to be the precursor of 2,3-dihydroxynaphthalene, a compound isolated by Fernley[149] from culture filtrates of anthracene-grown *P. aeruginosa.* 2,3-Dihydroxynaphthalene was considered to be a precursor of salicylate and catechol, but there is no evidence for this. Extracts of $P._{F}$ and $P._{G}$ apparently contained enzymes for the *meta*-cleavage of these catechols which were constitutively expressed. Catechol, 1,2-dihydroxynaphthalene, 1,2-dihydroxyanthracene and 3,4-dihydroxynaphthalene were all oxidized by extracts of cells grown with either naphthalene, anthracene, phenanthrene, or benzoate; it can be seen that much remains to be learned about the bacterial oxidation of anthracene and its analogs.

REFERENCES

1. **Stormer, K.**, Ueber die Wirkung des Schwefelkohlemstoffs und ahnlicher Stoffe auf den Boden. II, *Zentralbl. Bakteriol. Parasitenkd. Infektionskr.*, 20, 282, 1908.
2. **Zobell, C. E.**, Action of microorganisms on hydrocarbons, *Bacteriol. Rev.*, 10, 1, 1946.
3. **Treccani, V.**, Microbial degradation of hydrocarbons, *Prog. Ind. Microbiol.*, 4, 1, 1963.
4. **Van der Linden, A. C. and Thijsse, G. J. E.**, The mechanisms of microbial oxidations of petroleum hydrocarbons, *Adv. Enzymol.*, 27, 469, 1965.
5. **Gibson, D. T.**, The microbial oxidation of aromatic hydrocarbons, *Crit. Rev. Microbiol.*, 1, 199, 1971.
6. **Gibson, D. T.**, Initial reactions in the degradation of aromatic hydrocarbons, in *Biodegradation of Synthetic Organic Molecules in the Biosphere*, National Academy of Sciences, Washington, D.C., 1972, 116.
7. **Abbott, B. J. and Gledhill, W. E.**, The extracellular accumulation of metabolic products by hydrocarbon degrading microorganisms, *Adv. Appl. Microbiol.*, 14, 249, 1971.
8. **Jamison, V. M., Raymond, R. L., and Hudson, J. O.**, Hydrocarbon co-oxidation by *Nocardia corallina* V-49, *Dev. Ind. Microbiol.*, 12, 99, 1971.
9. **McKenna, E. J. and Kallio, R. E.**, The biology of hydrocarbons, *Ann. Rev. Microbiol.*, 19, 183, 1965.
10. **Subramanian, V., Sugumaran, M., and Vaidyanathan, C. S.**, Double hydroxylation reactions in microorganisms, *J. Ind. Inst. Sci.*, 60, 143, 1978.
11. **Perry, J. J.**, Microbial co-oxidations involving hydrocarbons, *Microbiol. Rev.*, 43, 59, 1979.
12. **Dagley, S.**, Catabolism of aromatic compounds by micro-organisms, *Adv. Microb. Physiol.*, 6, 1, 1971.
13. **Chapman, P. J.**, An outline of reaction sequences used for the bacterial degradation of phenolic compounds, in *Biodegradation of Synthetic Organic Molecules in the Biosphere*, National Academy of Sciences, Washington, D.C., 1972, 17.
14. **Dagley, S.**, Microbial degradation of organic compounds in the biosphere, *Am. Sci.*, 63, 681, 1975.
15. **Dagley, S.**, A biochemical approach to some problems of environmental pollution, *Essays Biochem.*, 11, 81, 1975.
16. **Dagley, S.**, Pathways for the utilization of organic growth substrates, in *The Bacteria*, Vol. 6, Ornston, L. N. and Sokatch, J. R., Eds., Academic Press, New York, 1979, 305.
17. **Ornston, L. N. and Stanier, R. Y.**, The β-ketoadipate pathway, *Adv. Microb. Physiol.*, 9, 89, 1973.
18. **Cerniglia, C. E. and Gibson, D. T.**, Metabolism of naphthalene by *Cunninghamella elegans*, *Appl. Environ. Microbiol.*, 34, 363, 1977.
19. **Cerniglia, C. E. and Gibson, D. T.**, Algal oxidation of aromatic hydrocarbons: formation of 1-naphthol from naphthalene by *Agmenellum quadruplicatum*, strain PR-6, *Biochem. Biophys. Res. Commun.*, 88, 50, 1979.
20. **Söhngen, N. L.**, Benzin, Petroleum, Paraffinol und Paraffin als. Kohlenstoff und Energiequelle fur Microben, *Zentralbl. Bakteriol. Parasitenkd. Infektionskr.*, Abt. II, 37, 595, 1913.
21. **Arnaudi, C., Canonica, L., and Treccani, V.**, Sulle Ossidazioni Microbiche di Idrocarburi e de Composite Aromatici, *Ric. Sci.*, 25, 3244, 1955.
22. **Kleinzeller, A. and Fencl, Z.**, Oxidation of benzene by a microorganism isolated from waste waters, *Chem. Listy.*, 46, 300, 1952; *Chem. Abstr.*, 47, 4209a, 1953.
23. **O'Connor, R. J., Weinrich, B. W., and Darlington, W. A.**, Phenol and the microbial conversion of benzene to catechol, *Bact. Proc.*, p. 97, 1964.
24. **Wieland, T., Griss, G., and Haccius, B.**, Untersuchungen zur mikrobiellen benzoloxydation, *Arch. Mikrobiol.*, 28, 383, 1958.
25. **Marr, E. K. and Stone, R. W.**, Bacterial oxidation of benzene, *J. Bacteriol.*, 81, 425, 1961.
26. **Gibson, D. T., Koch, J. R., and Kallio, R. E.**, Oxidative degradation of aromatic hydrocarbons. I. Enzymatic formation of catechol from benzene, *Biochemistry*, 7, 2653, 1968.
27. **Gibson, D. T., Cardini, G. E., Masales, F. C., and Kallio, R. E.**, Incorporation of oxygen-18 into benzene by *Pseudomonas putida*, *Biochemistry*, 9, 1631, 1970.
28. **Williams, R. T.**, *Detoxification Mechanisms*, 2nd ed., Chapman and Hall, London, 1959, 188.
29. **Daly, J. W., Jerina, D. M., and Witkop, B.**, Arene oxides and the NIH shift: the metabolism, toxicity and carcinogenicity of aromatic compounds, *Experientia*, 28, 1129, 1972.

30. **Jerina, D. M. and Daly, J. W.**, Arene oxides: a new aspect of drug metabolism, *Science*, 185, 173, 1974.
31. **Högn, T. and Jaenicke, L.**, Benzene metabolism of *Moraxella* species, *Eur. J. Biochem.*, 30, 369, 1972.
32. **Axcell, B. C. and Geary, P. J.**, Purification and properties of a soluble benzene-oxidizing system from a strain of *Pseudomonas*, *Biochem. J.*, 146, 173, 1975.
33. **Axcell, B. C. and Geary, P. J.**, The metabolism of benzene by bacteria, *Biochem. J.*, 136, 927, 1973.
34. **Ayengar, P. K., Hayaishi, O., Nakajima, M., and Tomida, I.**, Enzymic aromatization of 3,5-cyclohexadiene-1,2-diol, *Biochim. Biophys. Acta*, 33, 111, 1959.
35. **Kitagawa, M.**, Studies on the oxidation mechanism of methyl group, *J. Biochem.*, 43, 553, 1956.
36. **Reiner, A. M. and Hegeman, G. D.**, Metabolism of benzoic acid by bacteria. Accumulation of (−)-3,5-cyclohexadiene-1,2-diol-1-carboxylic acid by a mutant strain of *Alcaligenes eutrophus*, *Biochemistry*, 10, 2530, 1971.
37. **Claus, D. and Walker, N.**, The decomposition of toluene by soil bacteria, *J. Gen. Microbiol.*, 36, 107, 1964.
38. **Forro, J. R.**, The Microbial Oxidation of Toluene, Ph.D. dissertation, Pennsylvania State University, 1965.
39. **Ziffer, H., Jerina, D. M., Gibson, D. T., and Kobal, V. M.**, Absolute stereochemistry of the (+)-*cis*-1,2-dihydroxy-3-methylcyclohexa-3,5-diene produced from toluene by *Pseudomonas putida*, *J. Am. Chem. Soc.*, 95, 4048, 1973.
40. **Kobal, V. M., Gibson, D. T., Davis, R. E., and Garza, A.**, X-ray determination of the absolute stereochemistry of the initial oxidation product formed from toluene by *Pseudomonas putida* 39/D, *J. Am. Chem. Soc.*, 95, 4420, 1973.
41. **Gibson, D. T., Gschwendt, B., Yeh, W. K., and Kobal, V. M.**, Initial reactions in the oxidation of ethylbenzene by *Pseudomonas putida*, *Biochemistry*, 12, 1520, 1973.
42. **Omori, T., Jigami, Y., and Minoda, Y.**, Microbial oxidation of α-methylstyrene and β-methylstyrene, *Agric. Biol. Chem.*, 38, 409, 1974.
43. **Jigami, Y., Omori, T., and Minoda, Y.**, A new ring fission product suggestive of an unknown reductive step in the degradation of *n*-butylbenzene by *Pseudomonas*, *Agric. Biol. Chem.*, 38, 1751, 1974.
44. **Jigami, Y., Omori, T., and Minoda, Y.**, The degradation of isopropylbenzene and isobutylbenzene by *Pseudomonas* sp., *Agric. Biol. Chem.*, 39, 1781, 1975.
45. **Jigami, Y., Omori, T., and Minoda, Y.**, A new metabolic divergence in the degradation of isopropylbenzene and isobutylbenzene by *Pseudomonas* sp., *Agric. Biol. Chem.*, 43, 2001, 1979.
46. **Jigami, Y., Kawasaki, Y., Omori, T., and Minoda, Y.**, Coexistence of different pathways in the metabolism of *n*-propylbenzene by *Pseudomonas* sp., *Appl. Environ. Microbiol.*, 38, 783, 1979.
47. **Raymond, R. L., Jamison, V. W., and Hudson, J. O.**, Microbial hydrocarbon co-oxidation. I, *Appl. Microbiol.*, 15, 857, 1967.
48. **Metzler, D. E.**, *Biochemistry. The Chemical Reactions of Living Cells*, Academic Press, New York, 1977.
49. **Webley, D. M., Duff, R. B., and Farmer, V. C.**, Evidence for β-oxidation in the metabolism of saturated hydrocarbons by soil *Nocardias*, *Nature (London)*, 178, 1467, 1956.
50. **Davis, J. B. and Raymond, R. L.**, Oxidation of alkyl-substituted cyclic hydrocarbons by a *Nocardia* during growth on *n*-alkanes, *Appl. Microbiol.*, 9, 383, 1961.
51. **Sariaslani, F. S., Harper, D. B., and Higgins, J. J.**, Microbial degradation of hydrocarbons. Catabolism of 1-phenylalkanes by *Nocardia salmonicolor*, *Biochem. J.*, 140, 31, 1974.
52. **Seidman, M. M., Toms, A., and Wood, J. M.**, Influence of side-chain substituents on the position of cleavage of the benzene ring by *Pseudomonas fluorescens*, *J. Bacteriol.*, 97, 1192, 1969.
53. **Lunt, D. and Evans, W. C.**, The microbial metabolism of biphenyl, *Biochem. J.*, 118, 54P, 1970.
54. **Ahmed, M. and Focht, D. D.**, Degradation of polychlorinated biphenyls by two species of *Achromobacter*, *Can. J. Microbiol.*, 19, 47, 1973.
55. **Focht, D. D. and Alexander, M.**, Bacterial degradation of diphenylmethane, a DDT model substrate, *Appl. Microbiol.*, 20, 608, 1970.
56. **Focht, D. D. and Alexander, M.**, Aerobic cometabolism of DDT analogues by *Hydrogenomonas* sp., *J. Agric. Food Chem.*, 19, 20, 1971.
57. **Catelani, D., Sorlini, C., and Treccani, V.**, The metabolism of biphenyl by *Pseudomonas putida*, *Experientia*, 27, 1173, 1971.
58. **Gibson, D. T., Roberts, R. L., Wells, M. C., and Kobal, V. M.**, Oxidation of biphenyl by a *Beijerinckia* sp., *Biochem. Biophys. Res. Commun.*, 50, 211, 1973.
59. **Catelani, D., Colombi, A., Sorlini, C., and Treccani, V.**, Metabolism of biphenyl. 2-Hydroxy-6-oxo-6-phenylhex-2,4-dienoate: the *meta*-cleavage product from 2,3-dihydroxybiphenyl by *Pseudomonas putida*, *Biochem. J.*, 134, 1063, 1973.

60. **Catelani, D. and Colombi, A.**, Metabolism of biphenyl. Structure and physicochemical properties of 2-hydroxy-6-oxo-6-phenylhexa-2,4-dienoic acid, the *meta*-cleavage product from 2,3-dihydroxybiphenyl by *Pseudomonas putida*, *Biochem. J.*, 143, 431, 1974.
61. **Francis, A. J., Spanggord, R. J., Ouchi, G. I., Bramhall, R. J., and Bohonos, N.**, Metabolism of DDT analogues by a *Pseudomonas* sp., *Appl. Environ. Microbiol.*, 32, 213, 1976.
62. **Francis, A. J., Spanggord R. J., Ouchi, G. I., and Bohonos, N.**, Co-metabolism of DDT analogs by a *Pseudomonas* sp., *Appl. Environ. Microbiol.*, 35, 364, 1978.
63. **Kallio, R. E.**, in *Fermentation Advances*, D. Perlman, Ed., Academic Press, New York, 1969, 635.
64. **Baggi, G., Catelani, D., Galli, E., and Treccani, V.**, The microbial degradation of phenylalkanes. 2-Phenylbutane, 3-phenylpentane, 3-phenyldodecane and 4-phenylheptane, *Biochem. J.*, 126, 1091, 1972.
65. **McKenna, E. J.**, in *Biodegradation of Synthetic Organic Molecules in the Biosphere*, National Academy of Sciences, Washington, D.C., 1972, 73.
66. **Catelani, D., Colombi, A., Sorlini, C., and Treccani, V.**, Metabolism of quaternary carbon compounds: 2,2-dimethylheptane and tertbutylbenzene, *Appl. Environ. Microbiol.*, 34, 351, 1977.
67. **Lampel, K.**, The *p*-Cymene Pathway in *Pseudomonas putida*: A Comparative Study of Three Enzymes Catalysing Sequential Reactions Subsequent to Ring Cleavage, Ph.D. dissertation, University of Miami, Florida, 1980.
68. **Davis, R. S., Hosler, F. E., and Stone, R. W.**, Metabolism of *p*- and *m*-xylene by species of *Pseudomonas*, *Can. J. Microbiol.*, 14, 1005, 1968.
69. **Omori, T. and Yamada, K.**, Studies on the utilization of hydrocarbons by microorganisms. XIII, *Agric. Biol. Chem.*, 33, 979, 1969.
70. **Omori, T., Horiguchi, S., and Yamada, K.**, Studies on the utilization of hydrocarbons by microorganisms. X, *Agric. Bio. Chem.*, 31, 1337, 1967.
71. **Omori, T. and Yamada, K.**, Studies on the utilization of hydrocarbons by microorganisms. XV, *Agric. Biol. Chem.*, 34, 659, 1970.
72. **Omori, T. and Yamada, K.**, Studies on the utilization of hydrocarbons by microorganisms. XVII, *Agric. Biol. Chem.*, 34, 664, 1970.
73. **Dagley, S., Chapman, P. J., Gibson, D. T., and Wood, J. M.**, Degradation of the benzene nucleus by bacteria, *Nature (London)*, 202, 775, 1964.
74. **Sala-Trepat, J. M. and Evans, W. C.**, The metabolism of 2-hydroxymuconic semialdehyde by *Azotobacter* species, *Biochem. Biophys. Res. Commun.*, 43, 456, 1971.
75. **Catterall, F. A., Sala-Trepak, J. M., and Williams, P. A.**, The coexistence of two pathways for the metabolism of 2-hydroxymuconic semialdehyde in a naphthalene utilizing pseudomonad, *Biochem. Biophys. Res. Comm.*, 43, 463, 1971.
76. **Fujiwara, M., Golovleva, Y., Saeki, E., Nozaki, M., and Hayaishi, O.**, Extradiol cleavage of 3-substituted catechols by an intradiol dioxygenase, pyrocatechase, from a pseudomonad, *J. Biol. Chem.*, 250, 4848, 1975.
77. **Feist, C. F. and Hegeman, G. D.**, Phenol and benzoate metabolism by *Pseudomonas putida*; regulation of tangential pathways, *J. Bacteriol.*, 100, 869, 1969.
78. **Cain, R. B. and Farr, D. R.**, Metabolism of arylsulphonates by microorganisms, *Biochem. J.*, 106, 859, 1968.
79. **Murray, D., Duggleby, C. J., Sala-Trepat, J. M., and Williams, P. A.**, The metabolism of benzoate and methylbenzoates *via* the *meta*-cleavage pathway by *Pseudomonas arvilla* mt-2, *Eur. J. Biochem.*, 28, 301, 1972.
80. **Davey, J. F. and Gibson, D. T.**, Bacterial metabolism of *para*- and *meta*-xylenes: oxidation of a methyl substituent, *J. Bacteriol.*, 119, 923, 1974.
81. **Worsey, M. J. and Williams, P. A.**, Metabolism of toluene and the xylenes by *Pseudomonas putida* (arvilla) mt-2: evidence for a new function of the TOL plasmid, *J. Bacteriol.*, 124, 7, 1974.
82. **Reiner, A. M. and Hegeman, G. D.**, Metabolism of benzoic acid by bacteria. Accumulation of (−),3,5-Cyclohexadiene-1,2-diol-1-carboxylic acid by a mutant strain of *Alcaligenes eutrophus*, *Biochemistry*, 10, 2530, 1971.
83. **Nakazawa, T. and Yokota, T.**, Conjugal transfer of benzoate pathway genes in *Pseudomonas arvilla* mt-2 (in Japanese), *Jpn. J. Bacteriol.*, 28, 46, 1973.
84. **Williams, P. A. and Murray, K.**, Metabolism of benzoate and the methylbenzoates by *Pseudomonas putida (arvilla)* mt-2: evidence for the existence of a TOL plasmid, *J. Bacteriol.*, 120, 416, 1974.
85. **Wong, C. L. and Dunn, N. W.**, Transmissible plasmid coding for the degradation of benzoate and *m*-Toluate in *Pseudomonas arvilla* mt-2, *Genet. Res.*, 23, 227, 1974.
86. **Worsey, M. J. and Williams, P. A.**, Metabolism of toluene and xylenes by *Pseudomonas putida (arvilla)* mt-2: evidence for a new function of the TOL plasmid, *J. Bacteriol.*, 124, 7, 1975.
87. **Williams, P. A. and Worsey, M. J.**, Ubiquity of plasmids in coding for toluene and xylene metabolism in soil bacteria: evidence for the existence of new TOL plasmids, *J. Bacteriol.*, 125, 818, 1976.

88. **Friello, D. A., Mylroie, J. R., Gibson, D. T., Rogers, J. E., and Chakrabarty, A. M.**, XYL, a nonconjugative xylene degradative plasmid in *Pseudomonas* Pxy, *J. Bacteriol.*, 127, 1217, 1976.
89. **Kunz, D. A. and Chapman, P. J.**, Catabolism of pseudocumene and 3-ethyltoluene by *Pseudomonas putida (arvilla)* mt-2: evidence for new functions of the TOL(pWWO) plasmid, *J. Bacteriol.*, 146, 179, 1980.
90. **Kunz, D. A. and Chapman, P. J.**, Isolation and characterization of spontaneously occurring TOL plasmid mutants of *Pseudomonas putida* HS1, *J. Bacteriol.*, 146, 952, 1980.
91. **Kunz, D. A. and Ribbons, D. W.**, unpublished observations, 1980.
92. **Madhyastha, K. M. and Bhattacharyya, P. K.**, Microbiological transformations of terpenes. XII. Fermentation of *p*-cymene by a soil pseudomonad (PL strain), *Ind. J. Biochem.*, 5, 102, 1968.
93. **Madhyastha, K. M. and Bhattacharyya, P. K.**, Microbiological transformations of terpenes. XIII. Pathways for degradation of *p*-cymene in a soil pseudomonad (PL strain), *Ind. J. Biochem.*, 5, 161, 1968.
94. **Madhyastha, K. M., Rangachari, P. N., Raghabendra Rao, M., and Bhattacharyya, P. K.**, Microbiological transformations of terpenes. XV. Enzyme systems in the catabolism of *p*-cymene in PL strain, *Ind. J. Biochem.*, 5, 167, 1968.
95. **Omori, K., Horiguchi, S., and Takahashi, J.**, Studies on the utilization of hydrocarbons by microorganisms. VI. Screening of aromatic hydrocarbon-assimilating microorganisms and cumic acid formation from *p*-cymene, *Agric. Biol. Chem.* 29, 943, 1965.
96. **Leavitt, R. I.**, Microbial oxidation of hydrocarbons. Oxidation of *p*-isopropyltoluene by a *Pseudomonas* sp., *J. Gen. Microbiol.*, 49, 411, 1967.
97. **DeFrank, J. J. and Ribbons, D. W.**, The *p*-cymene pathway in *Pseudomonas putida* PL: isolation of a dihydrodiol accumulated by a mutant, *Biochem. Biophys. Res. Commun.*, 70, 1129, 1976.
98. **DeFrank, J. J. and Ribbons, D. W.**, *p*-Cymene pathway in *Pseudomonas putida:* initial reactions, *J. Bacteriol.*, 129, 1356, 1977.
99. **DeFrank, J. J. and Ribbons, D. W.**, *p*-Cymene pathway in *Pseudomonas putida*: ring cleavage of 2,3-dihydroxy-*p*-toluate and subsequent reactions, *J. Bacteriol.*, 129, 1365, 1977.
100. **Ribbons, D. W. and Senior, P. G.**, 2,3-Dihydroxybenzoate 3,4-oxygenase from *P. fluorescens*: oxidation of a substrate analogue, *Arch. Biochem. Biophys.*, 138, 557, 1970.
101. **Wigmore, G. J., Smith, A. J., and Ribbons, D. W.**, Metabolism of *para*-substituted benzoates and toluenes with "2,3-dihydroxybenzoates" as ring cleavage substrates, *Abstr. Annu. Meet. Am. Soc. Microbiol.*, K188, 1977.
102. **Wigmore, G. J. and Ribbons, D. W.**, *p*-Cymene pathway in *Pseudomonas putida*. Selective enrichment of defective mutants using halogenated analogs, *J. Bacteriol.*, 146, 920, 1980.
103. **Parke, D. V. and Williams, R. T.**, The hydroxylation of foreign aromatic compounds in the animal body, *Ann. Rep. Prog. Chem. (Chem. Soc. London)*, 55, 376, 1959.
104. **Booth, E. and Boyland, E.**, Formation of 1,2-dihydroxy-1,2-dihydronaphthalenes, *Biochem. J.*, 44, 361, 1949.
105. **Gibson, D. T., Ed.** *Detoxication of Environmental Pollutants*, Marcel Dekker, New York, in press.
106. **Tattersfield, F.**, The decomposition of naphthalene in the soil and the effect on its insecticidal action, *Ann. Appl. Biol.*, 15, 57, 1928.
107. **Gray, P. H. H. and Thornton, G. G.**, Soil bacteria that degrade certain aromatic compounds, *Zentralbl. Bakteriol. Parasitenkd. Infektionskr.*, 73, 74, 1928.
108. **Tausson, W. O.**, Naphthalin als Kohlenstoffquelle fur Bacterien, *Planta*, 4, 214, 1927.
109. **Walker, N. and Wiltshire, G. H.**, The breakdown of naphthalene by a soil bacterium, *J. Gen. Microbiol.*, 8, 273, 1953.
110. **Strawinski, R. J. and Stone, R. W.**, Biological oxidation of naphthalene, *Can. J. Microbiol.*, 1, 206, 1954.
111. **Treccani, V.**, Produzione di acido salicilico da naftalina per azione di uno schizomicete isolato da terreni petroliferi, *Ann. Microbiol. Enzimol.*, 5, 232, 1953.
112. **Treccani, V., Walker, N., and Wiltshire, G. H.**, The metabolism of naphthalene by soil bacteria, *J. Gen. Microbiol.*, 11, 341, 1954.
113. **Murphy, J. F. and Stone, R. W.**, The bacterial dissimilation of naphthalene, *Can. J. Microbiol.*, 1, 579, 1955.
114. **Fernley, H. N. and Evans, W. C.**, Oxidative metabolism of polycyclic hydrocarbons by soil pseudomonads, *Nature (London)*, 182, 373, 1958.
115. **Canonica, L., Fiecchi, A., and Treccani, E. V.**, Sui prodotti di ossidazione microbica di alcune naftaline sostituite, Instituto Lombardo (R. Sc.), 91, 119, 1957.
116. **Klausmeier, R. E. and Strawinski, R. J.**, Microbial oxidation of naphthalene, *J. Bacteriol.*, 73, 461, 1957.
117. **Griffiths, E. and Evans, W. C.**, A cell-free perhydroxylase system from soil pseudomonads, with activity on aromatic hydrocarbons, *Biochem. J.*, 95, 51, 1965.

118. **Jerina, D. M., Daly, J. W., Jeffrey, A. M., and Gibson, D. T.,** *cis*-1,2-Dihydroxy-1,2-dihydronaphthalene: a bacterial metabolite from naphthalene, *Arch. Biochem. Biophys.*, 142, 394, 1971.
119. **Jeffrey, A. M., Yeh, H. J. C., Jerina, D. M., Patel, T. R., Davey, J. F., and Gibson, D. T.,** Initial reactions in the oxidation of naphthalene by *Pseudomonas putida, Biochemistry*, 14, 575, 1975.
120. **Catterall, F. A., Murray, K., and Williams, P. A.,** The configuration of the 1,2-dihydroxy-1,2-dihydronaphthalene formed in the bacterial metabolism of naphthalene, *Biochim. Biophys. Acta*, 237, 361, 1971.
121. **Davies, J. I. and Evans, W. C.,** Oxidative metabolism of naphthalene by soil pseudomonads: the ring-fission mechanism, *Biochem. J.*, 91, 251, 1964.
122. **Williams, P. A., Murray, K., and Sala-Trepat, J. M.,** The coexistence of two metabolic pathways in the meta cleavage of catechol by *Pseudomonas putida* N.C.I.B. 10105, *Biochem. J.*, 20, 18, 1971.
123. **Catterall, F. A., Sala-Trepat, J. M., and Williams, P. A.,** The coexistence of two pathways for the metabolism of 2-hydroxymuconic semialdehyde in a naphthalene-grown *Pseudomonad, Biochem. Biophys. Res. Commun.*, 43, 463, 1971.
124. **Treccani, V., Galli, E., Catelani, D., and Sorlini, C.,** Induction of 1,2- and 2,3-diphenol oxygenases in *Pseudomonas desmolyticum, Z. Allg. Mikrobiol.*, 8, 65, 1968.
125. **Catterall, F. A. and Williams, P. A.,** The coexistence of two metabolic pathways in the *meta*-cleavage of catechol by *Pseudomonas putida, Biochem. J.*, 128, 88P, 1972.
126. **Shamsuzzaman, K. M. and Barnsley, E. A.,** The regulation of naphthalene metabolism in Pseudomonads, *Biochem. Biophys. Res. Commun.*, 60, 582, 1974.
127. **Williams, P. A., Catterall, F. A., and Murray, K.,** Metabolism of naphthalene, 2-methylnaphthalene, salicylate and benzoate by *Pseudomonas* P_G: regulation of tangential pathways, *J. Bacteriol.*, 124, 679, 1975.
128. **Azoulay, E.,** Regulation de la biosynthese et de l'activite des catechol-oxygenases chez *Pseudomonas, Bull. Soc. Franc. Physiol. Vegetale*, 12, 111, 1966.
129. **Shamsuzzaman, K. M. and Barnsley, E. A.,** The regulation of naphthalene oxygenase in pseudomonads, *J. Gen. Microbiol.*, 88, 193, 1974.
130. **Barnsley, E. A.,** The induction of the enzymes of naphthalene metabolism in pseudomonads by salicylate and 2-aminobenzoate, *J. Gen. Microbiol.*, 88, 193, 1975.
131. **Barnsley, E. A.,** Role and regulation of the *ortho* and *meta* pathways of catechol metabolism in pseudomonads metabolizing naphthalene and salicylate, *J. Bacteriol.*, 125, 404, 1976.
132. **Treccani, V. and Fiecchi, A.,** Ossidazione microbica della 2-metilnaftalina, *Atti. IX Congr. Naz. Microbiol., Palermo*, IX, 139, 1956.
133. **Treccani, V. and Fiecchi, A.,** Ossidazione microbica della 2-metilnaftalena, *Ann. Microbiol. Enzimol.*, 8, 36, 1958.
134. **Canonica, di L., Fiecchi, A., and Treccani, V.,** Sin prodotti di ossidazione microbica di alcune naftaline sostituite, *Istituto Lombardo (R. Sc.) B*, 91, 119, 1957.
135. **Rogoff, M. H. and Wender, I.,** Methylnaphthalene oxidations by pseudomonads, *J. Bacteriol.*, 77, 783, 1959.
136. **Baggi, G., Catelani, D., and Galli, E.,** Inibizione della syntesi della metapirocatecasi in *Pseudomonas desmolyticum* abituato a crescere su 5-metilsalicilato, *Ann. Microbiol. Enzimol.*, 18, 243, 1968.
137. **Colla, C., Biaggi, C., and Treccani, V.,** Chemica delle fermentazioni. Ricerche sul dativo microbico dell antracene e fenantrene. I. Formazione di acidi ossinaftoici, *Lincei R. Sci. Nat.*, 23, 66, 1957.
138. **Rogoff, M. H. and Wender, I.,** The microbiology of coal. I. Bacterial oxidation of phenanthrene, *J. Bacteriol.*, 73, 264, 1957.
139. **Colla, C., Fiecchi, A., and Treccani, V.,** Ricerche sul metabolismo ossidativo microbico dell' antracene e del fenantrene, *Ann. Microbiol. Enzimol.*, 9, 87, 1959.
140. **Rogoff, M. H.,** Chemistry of oxidation of polycyclic aromatic hydrocarbons by soil pseudomonads, *J. Bacteriol.*, 83, 998, 1962.
141. **Jerina, D. M., Selander, H., Yagi, H., Wells, M. C., Davey, J. F., Mahadevan, V., and Gibson, D. T.,** Dihydrodiols from anthracene and phenanthrene, *J. Am. Chem. Soc.*, 98, 5988, 1976.
142. **Koreeda, M., Akhtar, M. N., Boyd, D. R., Neill, J. D., Gibson, D. T., and Jerina, D. M.,** Absolute stereochemistry of *cis*-1,2-, *trans*-1,2-, and *cis*-3,4-dihydrodiol metabolites of phenanthrene, *J. Org. Chem.*, 43, 1023, 1978.
143. **Evans, W. C., Fernley, H. N., and Griffiths, E.,** Oxidative metabolism of phenanthrene and anthracene by soil pseudomonads. The ring fission mechanism, *Biochem. J.*, 95, 819, 1965.
144. **Ribbons, D. W.,** The microbiological degradation of aromatic compounds, *Ann. Rep. Chem. Soc.*, 62, 445, 1966.
145. **Kiyohara, H., Nagao, K., and Nomi, R.,** Degradation of phenanthrene through *o*-Phthalate by an *Aeromonas* sp., *Agric. Biol. Chem.*, 40, 1075, 1976.
146. **Kiyohara, H. and Nagao, K.,** Enzymatic conversion of 1-hydroxy-2-naphthoate in phenanthrene-grown *Aeromonas* sp. S45P1, *Agric. Biol. Chem.*, 41, 705, 1977.

147. **Kiyohara, H. and Nagao, K.**, The catabolism of phenanthrene and naphthalene by bacteria, *J. Gen. Microbiol.*, 105, 69, 1978.
148. **Rogoff, M. H. and Wender, I.**, 3-Hydroxy-2-naphthoic acid as an intermediate in bacterial dissimilation of anthracene, *J. Bacteriol.*, 74, 108, 1957.
149. **Fernley, H. N.**, Ph.D. Dissertation, University of Wales, Bangor, 1959.
150. **Akhtar, M. N., Boyd, D. R., Thompson, N. J., Koreeda, M., Gibson, D. T., Mahadevan, V., and Jerina, D. M.**, Absolute stereochemistry of the dihydroanthrancene-*cis*- and *trans*-1,2-diols produced from anthrancene by mammals and bacteria, *J. Chem. Soc., (Perkin)*, I, 2506, 1975.
151. **Evans, W. C. and Gibson, D. T.**, personal communication, 1980.

Chapter 4

GENETIC AND BIOCHEMICAL BASIS FOR MICROBIAL TRANSFORMATIONS AND DETOXIFICATION OF MERCURY AND MERCURIAL COMPOUNDS*

Simon Silver and Thomas G. Kinscherf

TABLE OF CONTENTS

* Original research in our laboratory was supported by grants from the National Science Foundation and the National Institutes of Health.

I. INTRODUCTION

The other chapters in this monograph concern microbial metabolism and detoxification of environmental pollutants introduced by man's industrial activities. Since mercury and organomercurials are natural components of the ecosphere, our consideration of the role of bacteria in the mercury cycle is somewhat different. While human activities have certainly affected local distributions and balances of the element, the process of mercury transformation would exist and occur in the absence of any human influence. Our goal in this review is to sketch an overall picture of the mercury cycle in the biosphere, to emphasize the role of microbial systems in this cycle, and to present what is known about the genetic and biochemical bases for the microbial transformations.

In an earlier review,[74] mercury transformations were examined at levels ranging from the global biosphere to that of a single bacterium. We will refer frequently to this review on questions that have basically remained unchanged over the intervening 4 years. Substantial progress has been made in understanding the molecular processes involved in systems for mercuric and organomercurial resistance in this time. These newer findings will be emphasized.

II. MERCURY IN THE BIOSPHERE

Despite its relatively low concentration in the environment, mercury is one of the better understood elements which undergo a complex cycle in the biosphere. Due both to its high toxicity as a pollutant and to the surprising frequency of microbial systems of mercurial transformation, mercury has been the subject of a large literature and extensive research efforts. As a result, we can construct what seems to be a fairly accurate overview of the sources and distribution of various mercurials in the ecosystem.

First, it is useful to know some very basic mercury chemistry. There are three valence states for inorganic mercury: Hg^0 (metallic mercury), Hg^+ (mercurous ion), and Hg^{2+} (mercuric ion). Metallic mercury is well known for its property of being a liquid at atmospheric temperatures and pressures; it is also virtually insoluble in water (K_s of about $10^{-7}M$) and has a relatively high vapor pressure. Inorganic mercurous and mercuric ions exist in equilibrium due to chemical dismutation: $2\ Hg^+ \rightleftharpoons Hg^0 + Hg^{2+}$.[46] Ionic forms of mercury are typically very reactive, readily forming inorganic and organic complexes; worth special mention are reactions with sulfhydryl-containing compounds to form stable mercaptides.

Since the mercury scare of the early 1970s, there has been a popular misconception that the introduction of mercury into the environment is a recent, man-made phenomenon. While it is certainly true that the polluting levels of mercury found in specific locales such as Minimata Bay, Japan, and the North American Great Lakes stem from human activities, it is important to realize that most of the mercury in the environment is not of human origin. A reasonable estimate is that human activities have added no more than 10^6 tons of mercury over the past century; this is contrasted to an estimated 2×10^8 tons of mercury dissolved in the oceans alone.[74] It is probable that most of the mercury of the earth lies beneath the crust, concentrating towards the core during planetary coalescence. A significant portion of crustal mercury is probably the product of volcanism and related seismic activities.

Mercury appears in a variety of rocks and soils, the highest concentrations being found in bedrock and deep-sea sediments.[40] The most abundant surface ore is cinnabar, a form of HgS.[14,20,40] The air above rocks and minerals rich in mercury contains high concentrations of the element; up to 16 ppb has been found in the air over mercury mines, with up to 1.2 ppb above undisturbed deposits of the metal ore.[20,82] Thus mercury can escape directly from the crust of the earth into the atmosphere. Rain can wash mercury out of the atmosphere, the rainwater reaching concentrations of 0.2 to 0.3 ppb (about 10^{-9} *M*). At the surface, rain can leach mercury from rocks and soils; eventually the mercury ends up associated with sulfides in the sediments of streams and lakes. The concentration of mercury downstream from mercury sources decreases rapidly (in one case by more than 3000 times over a distance of 50 km[82]) as the mercury complexes out of the stream and into the sediment. Mercury is found in seawater at levels between 0.1 and 1.2 ppb.

Mining and industrial activities add perhaps 12,500 metric tons of mercury to the environment each year. Mercury concentrations as high as 20,000 ppb have been found in coals and crude oils (this might be a hint of the availability of mercury in the prehistoric biosphere). Thus a large amount of mercury is released by burning fossil fuels, representing an estimated 10,000 to 60,000 tons per year.[33,74] Airborne Hg^0 and organomercurials are also introduced in notable quantities by urban sewage works,[68] undoubtedly due to microbial activity in the sewage.[74]

Mercury in the biosphere undergoes a cycle involving three predominant forms, Hg^0, Hg^{2+}, and CH_3Hg^+ (Figure 1). A major concern for us in this cycle comes from the finding that most of the mercury found in fish consumed by humans is methylmercury, a form 50 to 100 times more toxic than inorganic mercuric. Most atmospheric mercury is found as Hg^0 and CH_3Hg^+; surprisingly little, <1%, is found as dimethylmercury,[34,68] a more volatile form[90] believed to be a likely end-product in microbial transformations.[91] With established microbial systems for the production of both Hg^0 and CH_3Hg^+, it is very likely that microbes play a major role in the cycling of mercury in the biosphere. Microbes can also oxidize elemental mercury to ionic mercury,[29a] but the mechanism of this transformation is as yet unstudied.

III. MICROBIAL TRANSFORMATIONS OF MERCURY

A. Methylation of Mercury

Greater than 80% of the mercury found in fish (such as tuna and swordfish) is methylated, including that in fish living in uncontaminated off-shore waters.[32,89] The levels of mercury (presumably mostly methylmercury) found in deep-sea fish have not changed over the past 100 years,[3,45] making it unlikely that human activity is the source of this mercury. The primary form of mercury in both salt and fresh water is Hg^{2+};[9] therefore some process would seem to be at work to affect the transformation of mercury to methylmercury. Three alternatives have been proposed.

First, methylmercury production can be totally abiotic, without the intervention of microbes or microbial metabolites. Akagi and Takabatake[1] studied the photochemical methylation of Hg^{2+}. As much as 3%/day of mercuric acetate was converted to methylmercury by ultraviolet (UV) irradiation. This is at least two orders of magnitude more methylation than generally is found from microbial activities. Nevertheless, the photochemical methylation of Hg^{2+} was inhibited more than 99.9% by use of mercuric chloride plus acetic acid instead of mercuric acetate. This might preclude a role for photochemical methylation of Hg^{2+} in saline waters such as the sea.

Second, bacteria living in bottom sediment and sludge can carry out methylation by excreting methycobalamin, which serves as a methyl donor in vitro. Bottom-dwelling bacteria,[29,56,69,70] soil anaerobes,[79,94-96] and yeast[41] have the ability to methylate mercury, and the mechanism of nonenzymatic methylation via methylcobalamin is well documented.[5,13,93] Methylation from methylcobalamin is possibly the primary route

$$CH_3Hg^+ \rightleftharpoons Hg^{2+} \rightleftharpoons Hg^{\circ}$$

FIGURE 1. Interconversion of methyl- and inorganic mercury.

for methylation and mobilization of mercury from bottom sediments in polluted waters. As such, methylation in the sediments is of serious concern in the Great Lakes of North America and in the streams and rivers downstream from paper mills and chloralkali plants discharging large amounts of inorganic mercury.

A third possibility is that mercury may be methylated by the normal bacterial flora of the fish gill and gut or by fish organs directly.[57] Normal aerobic microorganisms as well as anaerobic sediment microbes can form methylmercury.[84] Even human intestinal bacteria have been shown to methylate Hg^{2+};[16,62] indeed, *Escherichia coli* K-12 was the best mercury methylater in one study by Brinckman and Iverson (as quoted in Silver et al.[66]). Wood and co-workers[61,91,92] have argued, on chemical grounds, that methylcobalamin is the natural methyl donor to mercury and that *S*-adenosylmethionine and N^5-methyltetrahydrofolate are incapable of donating methyl groups to Hg^{2+}; the methyl groups in the latter two cases would be transferred as CH_3^+, whereas methylcobalamin transfers CH_3^-. Nevertheless, *E. coli*, which does not produce methylcobalamin (but which can utilize it in methionine biosynthesis), does methylate Hg^{2+}.[66,84] Two alternative explanations remain for these findings. Either the *E. coli* (and other noncobalamin-synthesizing bacteria) are utilizing the trace levels of this vitamin that are found in the rich nutrient media used in all such experiments reported to date, or enzymatic transfer of methyl groups from alternative methyl donors occurs intracellularly. Wood[91] has emphasized that nonenzymatic direct transfer of CH_3 groups to Hg^{2+} cannot occur from *S*-adenosylmethionine or N^5-methyltetrahydrofolate; whether or not enzymatic transfer can occur still has not been tested to our knowledge. This question is still high on the list of important unanswered problems in our understanding of the biological mercury cycle.

B. Microbial Resistance to Mercury and Organomercurials

1. Occurrence

The earliest observation of the presence of bacteria with strong resistances toward mercury and organomercurials came from Japan, where resistant bacteria were isolated from organomercurial-polluted soil,[80,81,83] and from Europe, among hospital isolates of *Staphylococcus aureus*.[60,83] Subsequently mercury-resistant bacteria were isolated in the U.S., Japan, and elsewhere in surveys of isolates from hospital patients and from urban and industrial sewage. Many of the antibiotic-resistance plasmids in both Gram-negative and Gram-positive bacteria have the genes which determine resistance to Hg^{2+} and to organomercurials as well.[15,35,42,55,58,60,63,67,73] Indeed, both in Japan and in Australia plasmids conferring resistance to Hg^{2+} were perhaps twice as frequent among clinical isolates of *Pseudomonas aeruginosa* as were those conferring resistance to antibiotics, and about three quarters of the antibiotic-resistance plasmids also had markers for Hg^{2+} resistance. No naturally occurring, chromosomally determined resistance to mercury has been found, although the original mercury-resistant marine pseudomonads[8,51,52,85] have not been tested in this regard. Among hospital isolates of enteric bacteria, most were able to transfer resistance, implying that the determinants were plasmid-borne.[48] In a collection of over 800 plasmids transferred into a common host cell (*E. coli* K-12) from a wide range of species, including *Proteus, Providentia, Salmonella, Serratia, Shigella*, and *Klebsiella*, the frequency of genetic determinants of mercury resistance was about 25%.[63] That is approximately the percentage of Hg^{2+}-resistant isolates among hospital *E. coli* in St. Louis[63] or among clinical isolates of enteric organisms in Brazil.[15] In Japan, the frequency of Hg^{2+} resistance in enteric clinical isolates was more than 50%.[48,49] Among *Pseudomonas*[6,7,36] and *Staphylococcus*[27,28,49,55,60,86] plasmids, the frequency of determinants for mercury resistance was also very high. See Addendum for more recent ideas.

How did mercuric resistance determinants arise and why are they so ubiquitous? These important questions are ones to which we lack satisfactory answers. There are situations such as in Chesapeake Bay, where mercury-resistant bacteria are rare in the unpolluted areas of the lower Bay, but mercury-resistance frequencies exceed 90% in the bacteria found in polluted waters of the upper Bay in Baltimore Harbor.[9] In this specific instance, exposure to industrial pollution may have provided the direct selection for mercury-resistant bacteria. However, direct selection from recognizable sources would not seem to be an explanation for the difference in frequency of mercury resistance between clinical isolates and isolates from animals and animal products. During the last 3 decades, the extensive use of antibiotics in medicine and as food supplements for livestock has caused a dramatic increase in the frequency of antibiotic resistance.[17] Mercury resistance was found at very high levels with clinical *S. aureus* cultures in recent studies in Eastern Europe, while little mercury resistance (less than 1%) was found in the staphylococcal isolates from nonhospitalized humans and from nearby poultry, cattle, and sheep farms.[59,89a] Antibiotic-resistance plasmids occurred when animals were grown on antibiotic-supplemented feed but mercury resistance was not seen. *S. aureus* associated with foodstuffs showed a 45% frequency of resistance to one or more antibiotics, while only 2% of the staphylococci isolated from food showed mercury resistance.[59]

The frequent association of mercury resistance with antibiotic-resistance plasmids in clinical isolates is somewhat curious. The link between mercurials and the clinical setting is not pronounced. Formerly, mercury compounds were widely used as disinfectants. This dates to the early days of bacterial studies[37] and before that to medieval and ancient medicine.[14] Yet, the increase in frequency of mercury-resistant strains appears to be recent and occurred concomitantly with the increase in the frequency of antibiotic-resistance plasmids, not before that. There was no intermediate time during which mercury- (and other heavy metal) resistance plasmids abound, but when antibiotic-resistance plasmids were rare. Furthermore, in populations of people that have not been exposed to antibiotic treatment,[27,56] mercury resistance is rare.

2. Resistance Range

Not all plasmids in all bacterial species confer exactly the same range of resistances. In Gram-negative bacteria, resistance to the organomercurials merbromin and fluorescein mercuric acetate (Figure 2) is always associated with resistance to Hg^{2+}.[87] However, resistance to other organomercurials is only rarely found. A small subclass (perhaps 2%) of enteric plasmids that confer resistance to Hg^{2+} also confer resistance to the organomercurials phenylmercuric acetate and thimerosal. This subclass (referred to as "broad-spectrum"-resistance plasmids[63,87]) has only been found in the A-C, L, and H_2 incompatibility groupings of enteric plasmids. The L and H_2 plasmids tested were all found first in *Serratia marcescens* strains of clinical origin. Plasmids found in *P. aeruginosa* are similar to those in *E. coli* in that they fall into a "narrow-spectrum" subclass (conferring resistance to Hg^{2+}, merbromin, and fluorescein-mercuric acetate and slight resistance to *p*-hydroxymercuribenzoate), and a "broad-spectrum" subclass (conferring resistance to phenylmercuric acetate, thimerosal, methylmercuric, and ethylmercuric compounds, as well as to those compounds that define narrow-spectrum resistance[7,87]). In a few cases, where the same plasmid can be stably maintained in both *E. coli* and *P. aeruginosa*, the resistance to *p*-hydroxymercuribenzoate was found only in *P. aeruginosa*,[47a] indicating that the host cell background affects the expression of

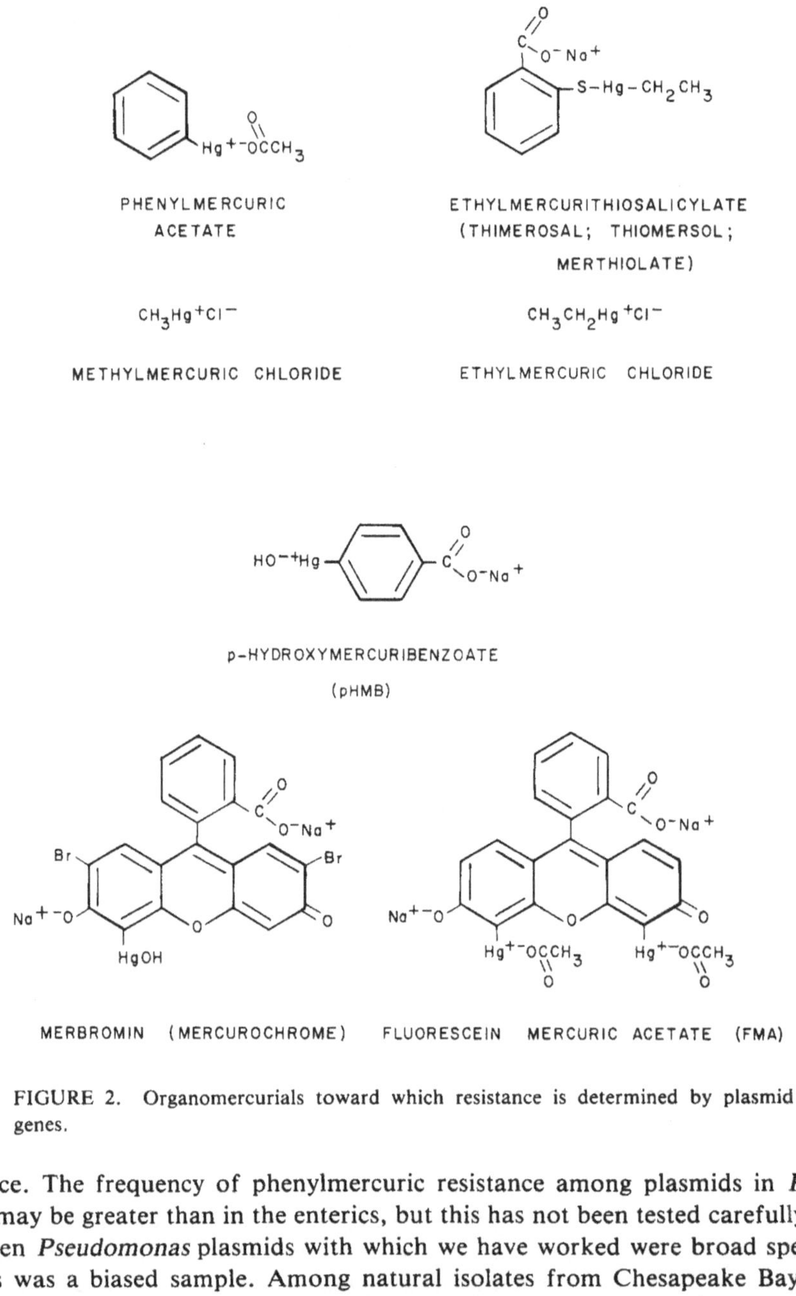

FIGURE 2. Organomercurials toward which resistance is determined by plasmid genes.

resistance. The frequency of phenylmercuric resistance among plasmids in *Pseudomonas* may be greater than in the enterics, but this has not been tested carefully. Four of the ten *Pseudomonas* plasmids with which we have worked were broad spectrum, but this was a biased sample. Among natural isolates from Chesapeake Bay, some mercury-resistant Pseudomonads were found to be broad spectrum in a small sampling.[8,9,51,52] With plasmid-bearing *S. aureus* strains, we have observed only the broad-spectrum class.[86,87] However, none of the eight plasmids tested conferred resistance to merbromin (unlike all of the more than 50 Hg^{2+}-resistance plasmids tested in *E. coli* and *Pseudomonas*). This complex pattern of resistance to organomercurials was summarized in greater detail recently.[87]

3. Chemistry of Mercury and Organomercurial Detoxification (Figure 3)

a. Hg^{2+} Detoxification

Three basic mechanisms of resistance to Hg^{2+} and organomercurials were considered by early investigators:[74] synthesis of mercury-binding thiols, permeability barriers lim-

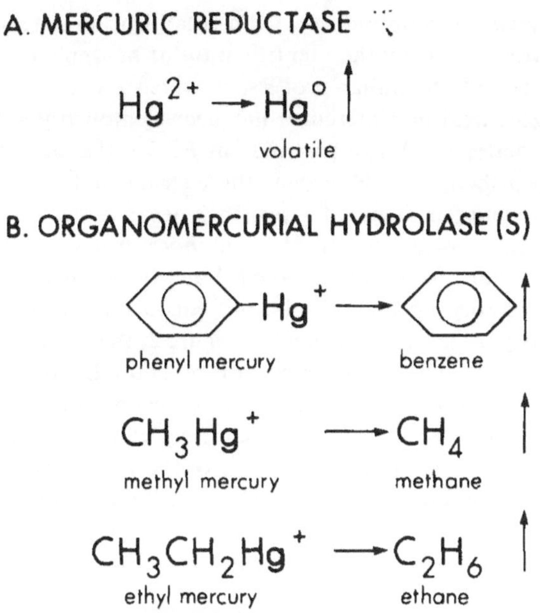

FIGURE 3. Enzymatic detoxification of Hg^{2+} and organomercurials.

iting access of the toxic material to the cell, and elimination of the toxic material from the growth medium. In every case of the more than 100 studied examples of bacterial resistance to Hg^{2+}, it was found that mercury was converted to a volatile form and released from the medium (Figure 3). No examples of plasmid-determined resistance to Hg^{2+} have appeared that do not involve volatilization.

Since both inorganic mercury and organomercurials can be volatile, the nature of the volatile mercury was a critical question. Furukawa et al.[22] determined directly that the volatile mercury was Hg^0 by collecting the product in a cold trap and visualizing the metallic liquid mercury formed. Subsequent isolation of the enzyme responsible for mercury volatilization in their soil isolate and determination of the activity[23,24] of this enzyme also indicated that reduction of Hg^{2+} to Hg^0 was required for volatilization. Komura and co-workers[38,39] also concluded that the volatilized mercury was Hg^0 from the observation that NADPH was oxidized in a Hg^{2+}-dependent reaction by an enzyme preparation from *E. coli.* With another Hg^{2+}-resistant *E. coli* strain, Summers and Silver[73] identified Hg^0 as the sole product of bacterial detoxification of Hg^{2+} by combined gas chromatography-mass spectroscopy (GC-MS). The product of bacterial action on $HgCl_2$ had a retention time in GC characteristic of Hg^0. The mass spectral pattern showed no sign of mercury-containing compounds of higher molecular weight. We have since considered microbially volatilized mercury and Hg^0 as synonymous. This is probably correct for the reactions determined by mercury-resistance plasmids, but it should be recalled that other volatile forms of mercury, e.g., CH_3HgCl, can result from bacterial activity in natural ecosystems.[34]

b. Detoxification of Organomercurials

The detoxification of phenylmercuric, ethylmercuric, and methylmercuric organomercurials is slightly more complicated, having essentially two steps. In the first step, mercury is cleaved from the organomercurial to yield benzene, ethane, and methane, respectively. Ethane and methane are volatile, as is, to a lesser extent, benzene. The second step is the reduction and subsequent volatilization of the released mercury. This process has been studied by examination of the volatilization reactions of radioactive (^{14}C and ^{203}Hg) substrates, and by the identification of benzene, ethane, and methane by gas chromatography.[23,76] A similar process, in which *p*-hydroxymercuribenzoate is cleaved to yield benzoic acid and mercury, has been demonstrated in an organomercurial-resistant soil isolate.[76,77] A few plasmids in *E. coli* also determine the ability to cleave *p*-hydroxymercuribenzoate. However, these strains did not show any increased resistance to *p*-hydroxymercuribenzoate, possibly because the decomposition is too slow to afford significant protection. In *P. aeruginosa*, broad-spectrum resistance includes resistance to *p*-hydroxymercuribenzoate, but without any evidence of chemical degradation;[7,87] this possibly represents a plasmid-encoded permeability barrier.

Of the other three organomercurials listed in Figure 2, the release of volatile mercury from thimerosal was found with broad-spectrum resistant *E. coli, P. aeruginosa*, and *S. aureus*. Resistance to fluoresceinmercuric acetate has been found with all mercury-resistant bacteria; resistance to merbromin has been found with all except *S. aureus*. There is no evidence for the degradation of either of these latter two organomercurials, once again raising the possibility of a permeability barrier mechanism of resistance.

4. Enzymology of Mercury and Organomercurial Detoxification

a. The Hg^{2+} Reductase

The enzyme responsible for the volatilization of mercury (mercuric reductase) has been purified from soil isolate K62 and three plasmid-containing *E. coli* strains. Crude enzyme preparations have been studied in preliminary ways from an additional 20 plasmid-bearing strains of *E. coli, P. aeruginosa*, and *S. aureus*. The results show a remarkable similarity among the enzymes, although some important differences are found.

In all tested cases, the mercuric reductase enzyme is found in the soluble protein fraction after cellular disruption;[24,38,64,75] there appears to be no membrane-bound form of the enzyme. In addition, the enzyme appears to be truly intracellular and not "periplasmic".[75] This seems a somewhat surprising location for a detoxifying enzyme, since the cell might be expected to keep the Hg^{2+} outside of the cell and away from potential target sites. However, with a reductive mechanism and the need for a small, easily diffusible co-factor, an intracellular location may be required. The co-factor for the plasmid-determined enzymes in *E. coli*,[38,64,75] *S. aureus*,[86,87] and *P. aeruginosa*[7,87] is NADPH. NADH was reported to substitute for NADPH for the enzyme of K62;[23,76] and in recent studies we have found that the degree of specificity depends upon the assay conditions.[56b] All tested mercuric reductases had a preference for NADPH, but could utilize NADH, given conditions of low Hg^{2+} and high NADH.

A most important difference appears to exist in the molecular weights of the mercuric reductase enzymes. Furukawa and Tonomura[24,25] estimate the molecular weight of the active mercuric reductase of K62 as 67,000 daltons from gel filtration studies. With purified *E. coli* enzymes, Schottel[64] found a molecular weight of approximately 180,000 daltons and determined that the protein was trimeric, with each of the polypeptide subunits having a molecular weight of near 67,000 daltons by SDS electrophoresis. The mercuric reductase from a *S. aureus* strain also appeared to be about 180,000 daltons from gel filtration measurements.[86] However, recently Fox and Walsh[19a] found that the enzyme from transposon Tn*501* is a dimer of molecular weight about 125,000 daltons and we have confirmed this finding.[36b] It appears tentatively as if the two immunological classes described below may represent trimeric and dimeric enzymes, all of which immunologically cross react.[36b]

All of the mercuric reductase enzymes share the properties of having a bound FAD moiety,[24,64] i.e., they are flavoproteins, and of requiring a thiol reagent for enzyme activity. The thiol reagent that acted preferentially varied from enzyme to enzyme (mercaptoethanol, cysteine, or thioglycolate), but in each case the requirement was for a reduced thiol at concentrations nearly 100-fold in excess of the substrate Hg^{2+}.[24,30,31,64]

The sulfhydryl compound is probably necessary to keep at least one sulfhydryl group on the enzyme in a reduced state.[75] However, the large molar excess of sulfhydryl (relative to the substrate Hg^{2+}) which is required for activity virtually assures that the form of mercury present will not be Hg^{2+} but $RSHg^{+}$ and possibly RSHgSR. The implication of this is that the substrate for the reductase(s) may not be ionic mercury but rather the thiol (or dithiol) adduct of mercury. A reaction sequence for the functioning of mercuric reductase has recently been proposed[19a] based on spectroscopic analysis. There are currently two alternative pathways, but both call for the addition of a mercuric-thiol adduct to an active site cysteine with the displacement of one of the mercury-bound soluble thiol moieties. The two pathways differ in the order of involvement of a second nearby active-site cysteine and of NADPH in reducing the enzyme-bound FAD to FAD_2.[19a]

Recently, a comparative immunological approach has provided added information about the degree of similarity of mercuric reductases from different plasmid sources. All mercuric reductases from Gram-negative bacteria that were tested had their enzyme activities inactivated by antiserum prepared against purified mercuric reductase determined by plasmid R831.[36b] These enzymes included those determined by plasmids from enteric organisms,[63] from *Pseudomonas*, from *Proteus*, and from a variety of soil and marine isolates. Another antiserum, prepared against the purified mercuric reductase determined by the transposon Tn*501*, distinguished at least three serotypes of enzyme in various Gram-negative bacterial extracts by "spur" formation in double immunodiffusion gels (Figure 4). Antibodies against the two Gram-negative enzymes failed to cross-react in any fashion with enzymes in extracts of three resistant strains of *S. aureus*. Thus although the Gram-positive and Gram-negative enzymes are of similar molecular weights and have similar co-factor requirements and kinetics, they are immunologically distinct.

This immunological distinction is reinforced by data on the relative thermostability of enzyme from Gram-negative sources as opposed to that of *S. aureus* enzyme. The Gram-negative enzymes were remarkably heat resistant, retaining most of their activity after 10 min at 80°C. The *S. aureus* enzymes were inactivated by temperatures about 25°C lower.[36b] In contrast, the phenylmercuric hydrolase activities of both Gram-positive and Gram-negative organisms were similarly temperature sensitive.

b. The Organomercurial Hydrolase

Somewhat less is known about the structure and function of the organomercurial hydrolases (or lyases) than is known about the Hg^{2+} reductase (Figure 3). It is clear, both with the soil isolate K62[76,77] and with *E. coli* plasmid-determined resistance,[64] that separate enzymes are involved in the hydrolytic cleavage and in Hg^{2+} reduction. When extracts of *E. coli* are passed through an affinity column (Blue Dextran), mercuric reductase is retained on the column while the hydrolase enzyme is not. After this single step, the hydrolase is free from contaminating reductase. Tezuka and Tonomura[76,77] demonstrated that two organomercurial hydrolase enzymes exist in their soil isolate, each with somewhat different properties and substrate ranges. Both organomercurial hydrolases have molecular weights of 19,000 from gel filtration measurements and have requirements for a reduced thiol reagent, preferably L-cysteine. The major enzyme activity has a substrate range that includes phenylmercuric, methylmercuric, ethylmercuric, and *p*-hydroxymercuribenzoate.[76] The second organomercurial hydrolase has a somewhat more limited substrate range of phenylmercuric and *p*-hydroxymercuribenzoate, but not methylmercuric.[77]

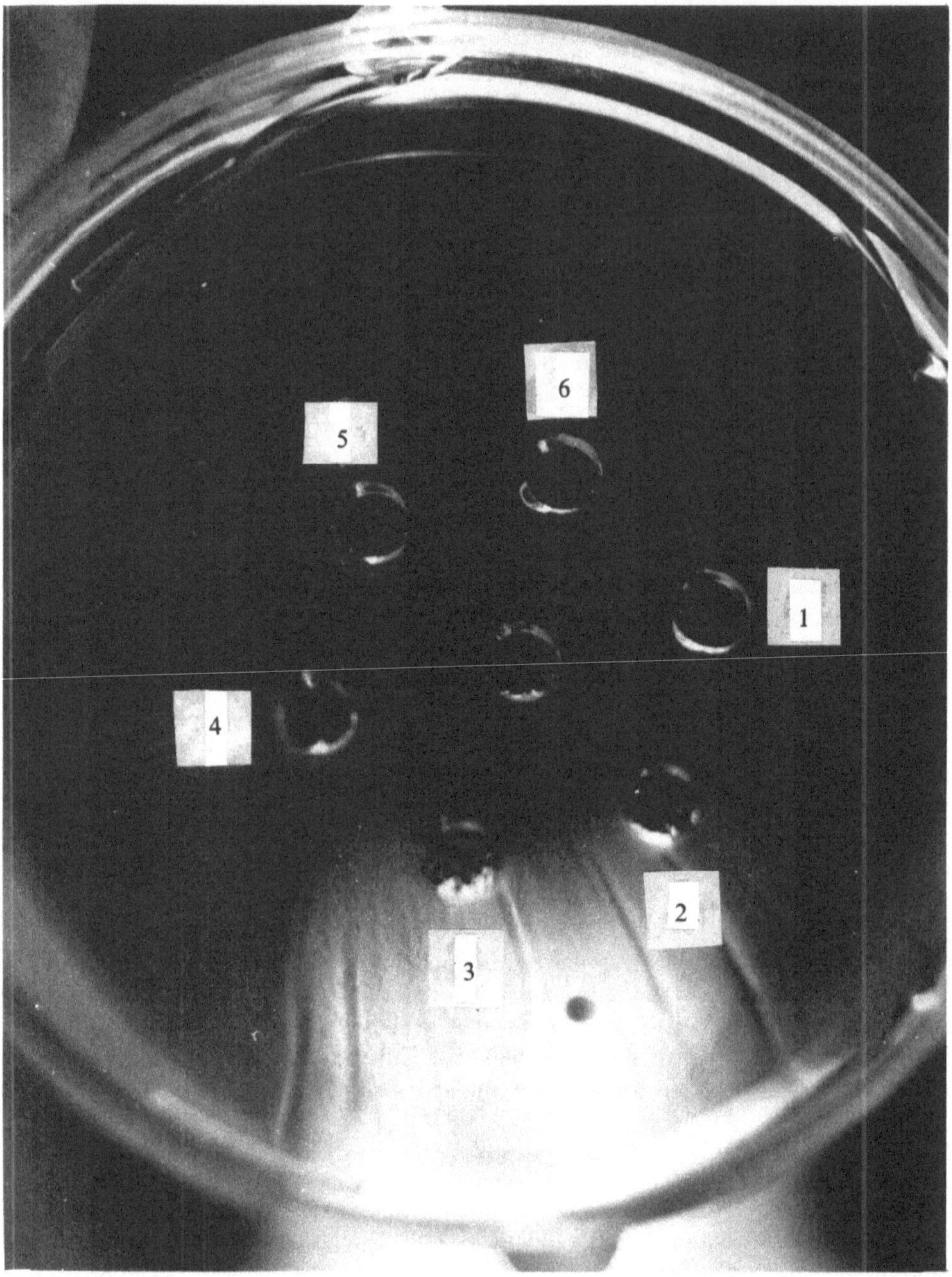

FIGURE 4. Double diffusion precipitant patterns with antibody against the mercuric reductase from transposon Tn*501* (center well) and mercuric reductase in crude extracts of *E. coli* containing plasmids. (1) JR211; (2) R28; (3) R938; (4) RIP135; (5) R455; (6) R52. See References 10 and 71 for plasmid characteristics.

Schottel[64] was not able to separate two organomercurial hydrolases from organomercurial-resistant *E. coli.* However, kinetic analysis showing two K_ms and two V_{max}s for phenylmercuric cleavage suggested the possibility of two enzymes. The molecular weight of the *E. coli* enzyme(s) appears to be somewhat larger than that reported for soil isolate K62, but the requirement for a thiol reagent remained.[64]

5. Induction and Transposition of mer

a. Inducibility

Mercury resistance is always inducible. The detoxifying enzymes are only synthesized by cells exposed to mercurials. There are no known examples of constitutive resistance to Hg^{2+} or organomercurials. The initial basis for this conclusion was that the addition of inhibitory concentrations of Hg^{2+} (or organomercurials) to sensitive or to plasmid-containing resistant cells caused rapid cessation of growth in both cases. After a delay of minutes or hours (depending on the amount of the mercurial added), the resistant cells resumed growth at a normal rate.[72] Previous growth on mercurials or Hg^{2+} prevents this initial growth delay.[85a] Hg^{2+} and the series of seven organomercurials we have studied (Figure 2) show complete cross-induction, in that induction by Hg^{2+} or any of the organomercurials confers resistance to all. The inducibility of mercurial resistance is paralleled by the inducibility of enzyme synthesis. Uninduced cells show no detectible mercuric reductase activity (less than 0.1% of the fully induced rate, e.g., Reference 19,) and no detectable organomercurial hydrolase activities. Varying the concentration of inducer increases volatilization of both mercuric salts and organomercurials, although not necessarily in a coordinate fasbion.[7,86]

b. The Transposability of the Genes Determining Mercury Resistance

It has been found recently that the resistance genes carried by plasmids often occur on a new class of genetic elements. These elements can move from plasmid to plasmid and from plasmid to chromosome, relocating genes in a process independent of normal host-cell recombination. The process is called transposition, and DNA segments that can accomplish transposition are called transposons. In some cases the genes determining mercury resistance (*mer*) occur on transposable genetic elements. Bennett and coworkers[4,71] characterized Tn*501*, a segment of DNA of 5.2 Mdaltons associated with the acquisition of the mercuric resistance phenotype from plasmid pVS1 of *P. aeruginosa.* The *mer* transposon from a pVS1 could insert into a minimum of five different sites on plasmid RP1.[26] Another transposon, Tn*A*, was inserted into the *mer* transposon[4] in order to study its functional substructure (see below). Friello and Chakrabarty[21] have recently shown that *mer* of *P. putida* plasmid MER is a 5.2-Mdalton transposon with identical restriction nuclease patterns to that of Tn*501*; tests in our laboratory show the enzyme of a MER strain to be of the same serotype as that of strains carrying Tn*501*.[36b] To date, there are no known differences between the transposon of MER strains and Tn*501*. However, another *mer* transposon, Tn*502*, has been found to be larger than Tn*501*,[70a] and codes for an enzyme of a different serotype.[36b] Other *mer* transposons recently have been reported.[2,75a] One of these, Tn*503*, is able to mobilize nontransmissible derivatives of the plasmid RP1.[2]

6. Mapping the mer Operon

a. Genetic Analysis of mer in S. aureus

Thus far, all mercury-resistant *S. aureus* strains express wide-spectrum type resistance. Resistances to Hg^{2+} and to phenylmercuric map closely together on the genetic maps of *S. aureus* plasmids pI258, pII147, and pI524, which are the three most thoroughly studied staphylococcal plasmids. Originally, the placement of these genes was based on deletion mutants that had lost both markers together. More recently,[43,47,53,54] a combination of deletion mapping, insertions of unrelated transposons leading to the loss of one or both resistances, restriction nuclease analysis, and heteroduplex DNA/DNA mapping has placed these markers more precisely on the plasmid maps. Lofdahl et al.[43] demonstrated that two restriction endonuclease *Eco*RI fragments were required in order to clone the *mer* region of plasmid pI258 into a small carrier plasmid. This was initial evidence for an *Eco*RI cleavage site within the *mer* operon. When the proximal portion of the pI258 was cloned alone, it gave a confused resistance pattern seen as "intermediate"[43] or a mixture of hypersensitive and fully resistant cells[65] depending upon the scoring procedure. The resistant cells made mercuric reductase enzyme under normal inducible conditions. This demonstrated that the regulatory region and at least most of the mercuric reductase gene were to be found on the proximal DNA fragment. Since phenylmercuric resistance[43] and volatilization of mercury from phenylmercuric[65] occurred only when both fragments were cloned together, one could conclude[65] that the gene for phenylmercuric hydrolase enzyme is either cleaved by the *Eco*RI enzyme or is to be found entirely on the distal fragment.

Novick et al.[53] concluded the same gene order based on transposon insertions into plasmid pI258. Some transposon insertions inactivated both Hg^{2+} and phenylmercuric resistances, while others (presumably more distal from the regulatory gene site) inactivated only phenylmercuric resistance, leaving mercuric resistance intact.[53] With an earlier series of 31 mercuric-sensitive mutants of *S. aureus*,[52a] some were found to lack both enzymes, while 9 produced essentially wild-type levels of phenylmercuric hydrolase and either zero or very low levels of mercuric reductase.[86] Thus either reductase alone or hydrolase alone could be lost by mutation, demonstrating the independent genetic loci for these enzymes. All of the Hg^{2+}-sensitive mutants were sensitive to phenylmercury, perhaps because reductase activity is required to volatilize Hg released by the hydrolase. Similar analysis with plasmids pII147 and pI524[47,54] have produced an identical ordering of the three genes involved: *merR* (the regulatory gene governing inducibility, *merA* (the structural gene for the mercuric reductase), and *merB* (the structural gene for the organomercurial hydrolase).

b. Analysis of the mer Operon of R100

When the transposon Tn*A*, an ampicillin-resistance determinant, was inserted into plasmid R100, a small percentage (2.5%) of all Tn*A* insertions were found to result in the loss of mercury resistance.[18,19] All 23 of these insertion mutants also became sensitive to merbromin and fluoresceinmercuric acetate (R100 codes for a narrow-spectrum-type resistance). The Tn*A* insertion mutants fell into four distinct classes.[19] (1) Five mutants (class A) produced no enzyme activity at all and were as sensitive to Hg^{2+} as the plasmidless strains; these were unable to complement one another or other Tn*A*-induced mutants. (2) Thirteen isolates produced no Hg^{2+} reductase under any tested condition; these mutant plasmids (class B) were capable of complementing class A mutants to bring about full resistance to Hg^{2+}. These class B mutants characteristically showed an increased binding of Hg^{2+} which could explain their hypersensitivity.[19] (3) Two sensitive mutants that constitutively produced low levels (2% of wild type) of mercuric reductase also were in complementation group B. (4) A second class of micro-constitutive mutants (class A above) could complement class B mutants to give rise to full Hg^{2+} resistance. Genetic complementation was seen between the Tn*A* transposon regulatory mutants of R100 and regulatory mutants in an unrelated plasmid, R828, indicating closely related control mechanisms.[19]

The results from these mutants enabled us to form the following model of the *mer* region (Figure 5). The regulatory gene, *merR*, is thought to govern synthesis of an inducer protein that controls RNA synthesis at an operator-promoter region in the *mer* operon. The reasons to postulate positive control include the finding of a mutant temperature sensitive for induction,[71a] and the observation that some insertion mutants of R100 have the expected properties for mutants in an inducer gene, that is low-level constitutive synthesis of the mercuric reductase. This level of synthesis could not be increased by induction.[19] These mutants were complemented by ColE1 plasmid carrying the cloned *Eco*RI restriction nuclease fragment H of plasmid R100,[19] indicating that the mutants are defective in a gene coded for by DNA in the H fragment (Figure

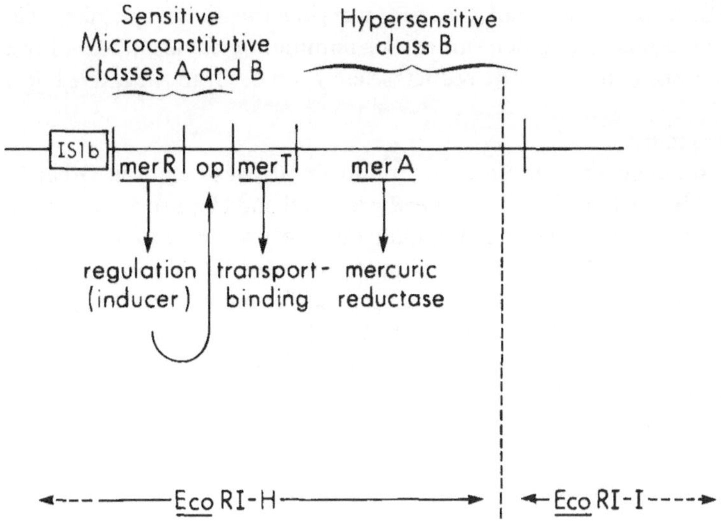

FIGURE 5. The genetic map of the *mer* operon of plasmid R100.

5). The other class mutants of R100 that are microconstitutive are postulated to have Tn*A* inserted into the operator region of the DNA, i.e., the receptor site for the inducer molecule. The uninduced wild-type plasmid showed 20 to 40 times less enzyme activity than the microconstitutive Tn*A* mutants. This suggests that the inducer protein may bind to the operator in the absence of Hg^{2+} and function negatively as a repressor, and then, in the presence of Hg^{2+}, functions positively to stimulate transcription of mRNA. As such, regulation of *mer* may be similar to regulation of the arabinose operon by the bifunctional *araC* protein. The mutants with hypersensitivity to Hg^{2+} are thought to have Tn*A* insertions in the Hg^{2+} reductase structural gene (*merA*; Figure 5) and their excess Hg^{2+} binding is ascribed to a mercury uptake function determined by a gene (*merT*) between the promoter region and the reductase gene.

Restriction nuclease and DNA cloning experiments on the *mer* region of R100 were done by Rownd and co-workers.[44,50] When *Eco*RI fragments H and I of plasmid R100 were cloned together (5.9 Mdalton total) into a carrier plasmid, *E. coli* carrying this plasmid showed normal narrow-spectrum mercury resistance. The *Eco*RI cleavage site between fragments H and I must lie within *mer*, as neither fragment conferred resistance if cloned without the other.[50] Cloned fragment *Eco*RI-H alone conferred a hypersensitivity phenotype, while *Eco*RI-I alone essentially had no effect (i.e., bacteria carrying cloned *Eco*RI-I by itself showed the same degree of mercury sensitivity as the nonresistant host *E. coli*). The hypersensitivity phenotype conferred by *Eco*RI-H was indistinguishable from that found with the hypersensitive Tn*A* insertion mutants, and *Eco*RI-H behaved the same way in complementation tests.[19] Initial kinetic and inhibitor studies indicate that hypersensitive bacteria bind more mercury than their sensitive counterparts, and lack the ability to detoxify (reduce) it.[50] Thus mercury hypersensitivity is thought to result from the expression of a transport-binding gene, *mer*T, coded entirely on *Eco*RI-H, in the absence of functional mercuric reductase, coded partially on *Eco*RI-I.

Sequence relationships between the *mer* region DNA of R100 and other plasmids have been studied[44a] using nick translation and Southern blot hybridization techniques. Plasmid R831, carrying wide-spectrum mercury-resistance genes, showed homology to both fragments *Eco*RI-H and *Eco*RI-I of R100. Plasmid R828, another wide-spectrum plasmid, also showed homology to the *mer* fragments of the narrow spectrum plasmid, as did ColE1::Tn*501*, a small plasmid carrying the mercury-resistance transposon Tn*501*. These results strengthen our earlier immunological arguments for a fair degree of similarity among the mercuric reductase enzymes of Gram-negative bacteria.

c. Sequencing *mer*

Recently, sequencing of the *mer* operon in Tn*501* was begun in Great Britain.[6a] Approximately 150 nucleotides of the *merR* proximal end (Figure 5) have been sequenced, including a Tn*A* mutation site; this indicates the known sequence extends into *mer*. The sequence includes presumptive initiation sites for RNA and protein synthesis. No constitutive RNA synthesis from this region was found in in vitro experiments.[5a] However, if *mer* in Tn*501* is similar to *mer* in R100 (Figure 5), then the first gene expressed would be *merR*, a regulatory gene whose expression would probably be limited. Most *mer* operon messenger RNA would be made only if the *merR* regulatory protein were present and activated.

7. Gene Products of the *mer* Operon

a. *mer* Transducing Phage

Insertion of the prophage λ into plasmid R100 gave rise to specialized λ *mer* transducing phages.[12] Mercuric reductase activity could be induced both during short-term infection and in stable transductants.[11] One of three independent λ *mer* isolates constitutively synthesized mercuric reductase during infection of nonlysogens, indicating transcription initiating at a λ promoter.[11] Analysis on SDS-polyacrylamide gels of *mer*-specific polypeptides produced after infection with λ *mer* showed three major products with molecular weights of 67,000 (presumably the mercuric reductase subunit), 11,700 and 8,800.[11] More recent studies[10] have confirmed the presence of three additional minor *mer*-related polypeptides with molecular weights of 52,400, 32,600, and 13,500.

b. Minicells

Further analysis of the gene products of the *mer* operon are currently underway. Preliminary experiments using "minicells" containing plasmid ColE1 with a Tn*501* insertion have yielded results that largely agree with those found with λ *mer*.[15a,36a] One interesting finding was that the 11,700-mol wt polypeptide could not be labeled with ^{35}S-cysteine; a lack of sulfhydryl groups might be beneficial to a protein directly exposed to mercury. Eventually, experiments with well-defined mutants and cloned *mer* fragments will permit the assignment of specific functions and cellular locales to the various *mer*-related polypeptides.

ACKNOWLEDGMENTS

We thank our many colleagues (referenced in the text) for exchanges of unpublished information and bacterial strains; also we would like to thank the typists of this manuscript, Ms. Linda Howard and Ms. Kim Persons. Recent research in this laboratory, described herein, was supported by grant PCM78-15787 from the National Science Foundation and AI15672 from the National Institutes of Health.

ADDENDUM

There has been considerable progress on several aspects of mercurial transformations since the initial writing of this Chapter. Here, we can only briefly list a few.

The question of what might be selecting mercurial resistance in hospital settings was puzzling a year ago. However, there is new evidence that use of mercurials in the hospitals is responsible for the high frequencies of mercurial resistance reported in this

Chapter. Porter et al.[97] found that the frequency of mercurial-resistant *S. aureus* had decreased to 2% by 1980 in St. Louis and ascribed that decrease to the reduced use of mercurials in hospital preparations, largely disinfectants and diuretics after the "mercury scare" of the early 1970s. During the late 1970s, the frequency of mercurial resistance in Tokyo hospital *S. aureus* also decreased by two thirds,[97] suggesting that we may expect this resistance to become less common in future years in hospital collections. Nevertheless, mercurial resistance was found in half of the *E. coli* isolatd from a Tokyo river in 1979—1980,[98] showing that environmental selection for mercurial resistance remains a strong force.

Recent progress in the assignment of specific *mer* polypeptides synthesized in minicells to regions of the operon and to specific gene functions has been accomplished by Jackson and Summers.[99] As expected, the large 67,000 dalton polypeptide represents the reductase subunit and was immunoprecipitated by antiserum prepared against mercuric reuctase. The antiserum also precipitated a 40,000 dalton polypeptide synthesized by the cloned *Eco*R1-H fragment (Figure 5). It appears as if the *Eco*R1 cleavage site indicated in the mercuric reductase *merA* gene in Figure 5 actually occurs about 60% into the gene, consistent with newer Tn*5* insertion mutant analysis of this region as well.[101] Two of the smaller *mer* specific peptides made in minicells appear to be associated with the bacterial inner membrane[100] and are thus candidates for the *merT* gene, which has thus far been hypothesized primarily from physiological studies. Jackson and Summer [99] may have a *merT* point mutant in a plasmid that confers reduced uptake of Hg^{2+}.

REFERENCES

1. **Akagi, H. and Takabatake, E.,** Photochemical formation of methylmercuric compounds from mercuric acetate, *Chemosphere,* 3, 131, 1973.
2. **Arwas, R. F. and Stanisich, V. A.,** The involvement of transposons in the mobilization of RP1 by the IncP2 plasmid pVS5 in *Pseudomonas aeruginosa,* unpublished manuscript, 1979.
3. **Barber, R. T., Vijayakumar, A., and Cross, F. A.,** Mercury concentrations in recent and ninety-year-old benthopelagic fish, *Science,* 178, 636, 1972.
4. **Bennett, P. M., Grinsted, J., Choi, C. L., and Richmond, M. H.,** Characterization of Tn *501,* a transposon determining resistance to mercuric ions, *Mol. Gen. Genet.,* 159, 101, 1978.
5. **Bertilsson, L. and Neujahr, H. Y.,** Methylation of mercury compounds by methylcobalamin, *Biochemistry,* 10, 2805, 1971.

5a. **Brown, N.L. Choi, C.L., Grinsted, J., Richmond, M.H., and Whitehead, P.R.,**Nucleotide sequences at the ends of the mercury resistance transposon, Tn*501, Nucleic Acid Res.,*8, 1933, 1980.

6. **Chakrabarty, A. M.,** Plasmids in *Pseudomonas, Ann. Rev. Genet.,* 10, 7, 1976.

6a. **Choi, C. L., Bennett, P. M., and Brown, N.,**personal communication.

7. **Clark, D. L., Weiss, A. A., and Silver, S.,** Mercury and organomercurial resistances determined by plasmids in *Pseudomonas, J. Bacteriol.,* 132, 186, 1977.
8. **Colwell, R. R. and Nelson, J. D., Jr.,** Metabolism of mercury compounds in microorganisms, EPA-600/3-75-007, Environmental Protection Agency, Washington, D.C., 1975.
9. **Colwell, R. R., Sayler, G. S., Nelson, J. D., Jr., and Justice, A.,** Microbial mobilization of mercury in the aquatic environment, in *Environmental Biogeochemistry: Metals Transfer and Ecological Mass Balances,* Vol. 2, Nriagu, J. O., Ed., Ann Arbor Science, Ann Arbor, 1976, 473.
10. **Dempsey, W. B. and McIntire, S. A.,** Lambda transducing phages derived from a FinO⁻R100::λ cointegrate plasmid: proteins encoded by the R100 replication/incompatibility region and the antibiotic resistance determinant, *Mol. Gen. Genet.,* 176, 319, 1979.
11. **Dempsey, W. B., McIntire, S. A., Willetts, N., Schottel, J., Kinscherf, T. G., Silver, S., and Shannon, W. A., Jr.,** Properties of lambda transducing bacteriophages carrying R100 plasmid DNA: mercury resistance genes, *J. Bacteriol.,* 136, 1084, 1978.
12. **Dempsey, W. B. and Willetts, N. S.,** Plasmid co-integrates of prophage lambda and R factor R100, *J. Bacteriol.,* 126, 166, 1976.

13. **DeSimone, R. E., Penley, M. W., Charbonneau, L., Smith, S. G., Wood, J. M., Hill, H. A. O., Pratt, J. M., Ridsdale, S., and Williams, R. J. P.**, The kinetics and mechanism of cobalamin-dependent methyl and ethyl transfer to mercuric ion, *Biochim. Biophys. Acta,* 304, 851, 1973.
14. **D'Itri, F. M.**, *The Environmental Mercury Problem,* CRC Press, Boca Raton, Fla., 1972.
14a. **D'Itri, P. A. and D'Itri, F. M.**, *Mercury Contamination: a Human Tragedy,* Interscience, New York, 1977.
15. **dos Reis, M. H. L., Ramos, M. J. C., and Trabulsi, L. R.**, Resistance to mercury in enteric organisms: frequency and genetic nature, *Rev. Microbiol.,* 9, 24, 1978.
15a. **Dougan, G.**, personal communication.
16. **Edwards, T. and McBride, B. C.**, Biosynthesis and degradation of methylmercury in human faeces, *Nature (London),* 253, 462, 1975.
17. **Falkow, S.**, *Infectious Multiple Drug Resistance,* Pion, London, 1975.
18. **Foster, T. J. and Nakahara, H.**, Deletions in the r-determinant *mer* region of plasmid R100-1 selected for loss of mercury hypersensitivity, *J. Bacteriol.,* 140, 301, 1979.
19. **Foster, T. J., Nakahara, H., Weiss, A. A., and Silver, S.**, Transposon *A*-generated mutations in the mercuric resistance genes of plasmid R100-1, *J. Bacteriol.,* 140, 167, 1979.
19a. **Fox, B., and Walsh, C.T.**, *J. Biol. Chem.,*in press, 1982.
20. **Friberg, L. and Vostal, J., Eds.**, *Mercury in the Environment,* CRC Press, Boca Raton, Fla., 1972.
21. **Friello, D. A. and Chakrabarty, A. M.**, Transposable mercury resistance in *Pseudomonas putida,* in *Plasmids and Transposons: Environmental Effects and Maintenance Mechanisms,* Stuttard, C. and Rozee, K. R., Eds., Academic Press, New York, 1980, 249.
22. **Furukawa, K., Suzuki, T., and Tonomura, K.**, Decomposition of organic mercurial compounds by mercury-resistant bacteria, *Agric. Biol. Chem.,* 33, 128, 1969.
23. **Furukawa, K. and Tonomura, K.**, Enzyme system involved in the decomposition of phenyl mercuric acetate by mercury-resistant *Pseudomonas, Agric. Biol. Chem.,* 35, 604, 1971.
24. **Furukawa, K. and Tonomura, K.**, Metallic mercury-releasing enzyme in mercury-resistant *Pseudomonas, Agric. Biol. Chem.,* 36, 217, 1972.
25. **Furukawa, K. and Tonomura, K.**, Induction of metallic mercury-releasing enzyme in mercury-resistant *Pseudomonas, Agric. Biol. Chem.,* 36, 2441, 1972.
26. **Grinsted, J., Bennett, P. M., Higginson, S., and Richmond, M. H.**, Regional preference of insertion of Tn*501* and Tn*802* into RP1 and its derivatives, *Mol. Gen. Genet.,* 166, 313, 1978.
27. **Groves, D. J., Short, H., Thewaini, A. J., and Young, F. E.**, Epidemiology of antibiotic and heavy metal resistance in bacteria: resistance patterns in *Staphylococci* isolated from populations in Iraq exposed and not exposed to heavy metals or antibiotics, *Antimicrob. Agents Chemother.,* 7, 622, 1975.
28. **Groves, D. J. and Young, F. E.**, Epidemiology of antibiotic and heavy metal resistance in bacteria: resistance patterns in *Staphylococci* isolated from populations not known to be exposed to heavy metals, *Antimicrob. Agents Chemother.,* 7, 614, 1975.
29. **Hamdy, M. K. and Noyes, O. R.**, Formation of methyl mercury by bacteria, *Appl. Microbiol.,* 30, 424, 1975.
29a. **Holm, H.W. and Cox, M.F.**, Transformation of elemental mercury by bacteria, *Appl. Microbiol.*29, 491, 1975.
30. **Izaki, K.**, Enzymatic reduction of mercurous ions in *Escherichia coli* bearing R factor, *J. Bacteriol.,* 131, 696, 1977.
31. **Izaki, K., Tashiro, Y., and Funaba, T.**, Mechanism of mercuric chloride resistance in microorganisms. III. Purification and properties of a mercuric ion reducing enzyme from *Escherichia coli* bearing R factor, *J. Biochem.,* 75, 591, 1974.
32. **Jernelöv, A. and Martin, A.-L.**, Ecological implications of metal metabolism by microorganisms, *Ann. Rev. Microbiol.,* 29, 61, 1975.
33. **Joensuu, O. I.**, Fossil fuels as a source of mercury pollution, *Science,* 172, 1027, 1971.
34. **Johnson, D. L. and Braman, R. S.**, Distribution of atmospheric mercury species near ground, *Environ. Sci. Technol.,* 8, 1003, 1974.
35. **Joly, B. and Cluzel, R.**, The role of heavy metals and their derivatives in the selection of antibiotics resistant gram-negative rods, *Ann. Inst. Pasteur Paris,* 126B, 51, 1975.
36. **Joly, B., Cluzel, R., Henri, Ph., and Barjot, J.**, The resistance of *Pseudomonas* to antibiotics and heavy metal: minimal inhibitory concentrations and genetic transfers, *Ann. Inst. Pasteur Paris,* 127B, 57, 1976.
36a. **Kinscherf, T. G.**, unpublished data.
36b. **Kinscherf, T. G. and Silver, S.**, in preparation.
37. **Koch, R.**, Ueber Disinfection, in *Kaiserlichen Gesundreitsamte,* Berlin, 1881, 234.
38. **Komura, I., Funaba, T., and Izaki, K.**, Mechanism of mercuric chloride resistance in microorganisms. II. NADPH-dependent reduction of mercuric chloride and vaporization of mercury from mercuric chloride by a multiple drug resistant strain of *Escherichia coli, J. Biochem.,* 70, 895, 1971.

39. **Komura, I. and Izaki, K.**, Mechanism of mercuric chloride resistance in microorganisms. I. Vaporization of a mercury compound from mercuric chloride by multiple drug resistant strains of *Escherichia coli*, *J. Biochem.*, 70, 885, 1971.
40. **Kothny, E. L.**, The three-phase equilibrium of mercury in nature, in *Trace Elements in the Environment*, Kothny, E. L., Ed., American Chemical Society, Washington, D.C., 1973, 48.
41. **Landner, L.**, Biochemical model for the biological methylation of mercury suggested from methylation studies *in vivo* with *Neurospora crassa*, *Nature (London)*, 230, 452, 1971.
42. **Laub-Kupersztejn, R., Thomas, J., and Pohl, P.**, Mercury and antibiotic resistance in *Enterobacteriaceae*, *Ann. Inst. Pasteur Paris*, 125B, 501, 1974.
43. **Löfdahl, S., Sjöström, J.-E., and Philipson, L.**, A vector for recombinant DNA in *Staphylococcus aureus*, *Gene*, 3, 161, 1978.
44. **Miki, T., Easton, A. M., and Rownd, R. H.**, Mapping of the resistance genes of the R plasmid NR1, *Mol. Gen. Genet.*, 158, 217, 1978.
44a. **Miller, J.R. and Silver, S.**, manuscript in preparation.
45. **Miller, G. E., Grant, P. M., Kishore, R., Steinkruger, F. J., Rowland, F. S., and Guinn, V. P.**, Mercury concentrations in museum specimens of tuna and swordfish, *Science*, 175, 1121, 1972.
46. **Moser, H. C. and Voigt, A. F.**, Dismutation of the mercurous dimer in dilute solutions, *J. Am. Chem. Soc.*, 79, 1837, 1957.
47. **Murphy, E. and Novick, R. P.**, Physical mapping of *Staphylococcus aureus* penicillinase plasmid pI524: characterization of an invertible region, *Mol. Gen. Genet.*, 175, 19, 1979.
47a. **Nakahara, H.**, unpublished.
48. **Nakahara, H., Ishikawa, T., Sarai, Y., Kondo, I., Kozukue, H., and Mitsuhashi, S.**, Mercury resistance and R plasmids in *Escherichia coli* isolated from clinical lesions in Japan, *Antimicrob. Agents Chemother.*, 11, 999, 1977.
49. **Nakahara, H., Ishikawa, T., Sarai, Y., Kondo, I., and Mitsuhashi, S.**, Frequency of heavy metal resistance in bacteria from inpatients in Japan, *Nature (London)*, 266, 165, 1977.
50. **Nakahara, H., Silver, S., Miki, T., and Rownd, R. H.**, Hypersensitivity to Hg^{2+} and hyperbinding activity associated with cloned fragments of the mercurial resistance operon of plasmid NR1, *J. Bacteriol.*, 140, 161, 1979.
51. **Nelson, J. D., Blair, W., Brinckman, F. E., Colwell, R. R., and Iverson, W. P.**, Biodegradation of phenylmercuric acetate by mercury-resistant bacteria, *Appl. Microbiol.*, 26, 321, 1973.
52. **Nelson, J. D., Jr. and Colwell, R. R.**, The ecology of mercury-resistant bacteria in Chesapeake Bay, *Microb. Ecol.*, 1, 191, 1975.
52a. **Novick, R. P.**, personal communication.
53. **Novick, R. P., Edelman, I., Schwesinger, M. D., Gruss, A. D., Swanson, E. C., and Pattee, P. A.**, Genetic translocation in *Staphylococcus aureus*, *Proc. Natl. Acad. Sci. U.S.A.*, 76, 400, 1979.
54. **Novick, R. P., Murphy, E., Gryczan, T. J., Baron, E., and Edelman, I.**, Penicillinase plasmids of *Staphylococcus aureus*: restriction-deletion maps, *Plasmid*, 2, 109, 1979.
55. **Novick, R. P. and Roth, C.**, Plasmid-linked resistance to inorganic salts in *Staphylococcus aureus*, *J. Bacteriol.*, 95, 1335, 1968.
56. **Olarte, J.**, personal communication.
56a. **Olson, B. H. and Cooper, R. C.**, Comparison of aerobic and anaerobic methylation of mercuric chloride by San Francisco Bay sediments, *Water Res.*, 10, 113, 1976.
56b. **Olson, G. J., Rubinstein, J., and Silver, S.**, manuscript in preparation.
57. **Pan, S.-K., Imura, N., and Ukita, T.**, Fractionation and characterization of mercury-methylation factor in tuna liver, *Chemosphere*, 2, 247, 1973.
58. **Pinney, R. J.**, Survival of plasmid-containing strains of *Escherichia coli*, *Pseudomonas aeruginosa* and *Staphylococcus aureus* in phenylmercuric nitrate and thiomersal, *J. Pharm. Pharmac.*, 30, 228, 1978.
59. **Raška, K., Matějovská, V., Matějovská, D., Bergdoll, M. S., and Petráš**, To the origin of contamination of foodstuff by enterotoxigenic staphylococci, in *Proc. 4th Int. Symp. Staphylococci and Staphylococcal Infections*, Jeljaszewicz, J., Ed., Gustav Fischer Verlag, Stuttgart, 381, 1981.
60. **Richmond, M. H. and John, M.**, Co-transduction by a staphylococcal phage of the genes responsible for penicillinase synthesis and resistance to mercury salts, *Nature (London)*, 202, 1360, 1964.
61. **Ridley, W. P., Dizikes, L. J., and Wood, J. M.**, Biomethylation of toxic elements in the environment, *Science*, 197, 329, 1977.
62. **Rowland, I. R., Grasso, P., and Davies, M. J.**, The methylation of mercuric chloride by human intestinal bacteria, *Experientia*, 31, 1064, 1975.
63. **Schottel, J., Mandal, A., Clark, D., Silver, S., and Hedges, R. W.**, Volatilisation of mercury and organomercurials determined by inducible R factor systems in enteric bacteria, *Nature (London)*, 251, 335, 1974.
64. **Schottel, J. L.**, The mercuric and organomercurial detoxifying enzymes from a plasmid-bearing strain of *Escherichia coli*, *J. Biol. Chem.*, 253, 4341, 1978.

65. **Silver, S.,** Mechanisms of plasmid-determined resistance to mercury, cadmium and arsenic in *Staphylococcus aureus,* in *Proc. 4th Int. Symp. Staphylococci and Staphylococcal Infections,* Jeljaszewicz, J., Ed., Gustav Fischer Verlag, Stuttgart, 589, 1981.
66. **Silver, S., Schottel, J., and Weiss, A.,** Bacterial resistance to toxic metals determined by extrachromosomal R factors, in *Proc. 3rd Int. Biodegradation Symp.,* Sharpley, J. M. and Kaplan, A. M., Eds., Applied Science, London, 1976, 899.
67. **Smith, D. H.,** R factors mediate resistance to mercury, nickel, and cobalt, *Science,* 156, 1114, 1967.
68. **Soldano, B. A., Bien, P., and Kwan, P.,** Air-borne organo-mercury and elemental mercury emissions with emphasis on central sewage facilities, *Atmos. Environ.,* 9, 941, 1975.
69. **Spangler, W. J., Spigarelli, J. L., Rose, J. M., Flippin, R. S., and Miller, H.,** Degradation of methylmercury by bacteria isolated from environmental samples, *Appl. Microbiol.,* 25, 488, 1973.
70. **Spangler, W. J., Spigarelli, J. L., Rose, J. M., and Miller, H. M.,** Methylmercury: bacterial degradation in lake sediments, *Science,* 180, 192, 1973.
70a. **Stanisich, V. A.,** personal communication.
71. **Stanisich, V. A., Bennett, P. M., and Richmond, M. H.,** Characterization of a translocation unit encoding resistance to mercuric ions that occurs on a nonconjugative plasmid in *Pseudomonas aeruginosa, J. Bacteriol.,* 129, 1227, 1977.
71a. **Summers, A. O.,** unpublished.
72. **Summers, A. O., Schottel, J., Clark, D., and Silver, S.,** Plasmid-borne Hg(II) and organomercurial resistance, in *Microbiology — 1974,* Schlessinger, D., Ed., American Society for Microbiology, Washington, D.C., 1975, 219.
73. **Summers, A. O. and Silver, S.,** Mercury resistance in a plasmid-bearing strain of *Escherichia coli, J. Bacteriol.,* 112, 1228, 1972.
74. **Summers, A. O. and Silver, S.,** Microbial transformations of metals, *Ann. Rev. Microbiol.,* 32, 637, 1978.
75. **Summers, A. O. and Sugarman, L. I.,** Cell-free mercury (II)-reducing activity in a plasmid-bearing strain of *Escherichia coli, J. Bacteriol.,* 119, 242, 1974.
75a. **Summers, A. O., Weiss, R. B., and Jacoby, G. A.,** Transposition of mercury resistance from a transferable R plasmid of *Escherichia coli, Plasmid,* 3, 35, 1980.
76. **Tezuka, T. and Tonomura, K.,** Purification and properties of an enzyme catalyzing the splitting of carbon-mercury linkages from mercury-resistant *Pseudomonas* K-62 strain. I. Splitting enzyme 1, *J. Biochem.,* 80, 79, 1976.
77. **Tezuka, T. and Tonomura, K.,** Purification and properties of a second enzyme catalyzing the splitting of carbon-mercury linkages from mercury-resistant *Pseudomonas* K-62, *J. Bacteriol.,* 135, 138, 1978.
78. **Timoney, J. F., Port, J., Giles, J., and Spanier, J.,** Heavy-metal and antibiotic resistance in the bacterial flora of sediments of New York Bight, *Appl. Environ. Microbiol.,* 36, 465, 1978.
79. **Tonomura, K., Furukawa, K., and Yamada, M.,** Microbial conversion of mercury compounds, in *Environmental Toxicology of Pesticides,* Matsumura, F., Ed., Academic Press, New York, 1972, 115.
80. **Tonomura, K. and Kanzaki, F.,** The reductive decomposition of organic mercurials by cell-free extract of a mercury-resistant pseudomonad, *Biochim. Biophys. Acta,* 184, 227, 1969.
81. **Tonomura, K., Nakagami, T., Futai, F., and Maeda, K.,** Studies on the action of mercury-resistant microorganisms on mercurials. I. The isolation of mercury-resistant bacterium and the binding of mercurials to the cells, *J. Ferment. Technol.,* 46, 506, 1968.
82. Mercury in the Environment, U.S. Geological Survey Professional Paper No. 713, U.S. Government Printing Office, Washington, D.C., 1970.
83. **Váczi, L., Fodor, M., Milch, H., and Réthy, A.,** Studies on the mercuric chloride resistance of *Staphylococcus aureus, Acta Microbiol. Pol.,* 9, 81, 1962.
84. **Vonk, J. W. and Sijpesteijn, A. K.,** Studies on the methylation of mercuric chloride by pure cultures of bacteria and fungi, *Antonie van Leeuwenhoek,* 39, 505, 1973.
85. **Walker, J. D. and Colwell, R. R.,** Mercury-resistant bacteria and petroleum degradation, *Appl. Microbiol.,* 27, 285, 1974.
85a. **Weiss, A. A.,** unpublished.
86. **Weiss, A. A., Murphy, S. D., and Silver, S.,** Mercury and organomercurial resistances determined by plasmids in *Staphylococcus aureus, J. Bacteriol.,* 132, 197, 1977.
87. **Weiss, A. A., Schottel, J. L., Clark, D. L., Beller, R. G., and Silver, S.,** Mercury and organomercurial resistance with enteric, staphylococcal, and pseudomonad plasmids, in *Microbiology — 1978,* Schlessinger, D., Ed., American Society for Microbiology, Washington, D.C., 1978, 121.
88. **Weiss, H. V., Koide, M., and Goldberg, E. D.,** Mercury in a Greenland sheet: evidence of recent input by man, *Science,* 174, 692, 1971.
89. **Westöö, G.,** Methylmercury as percentage of total mercury in flesh and viscera of salmon and sea trout of various ages, *Science,* 181, 567, 1973.

89a. **Witte, W.**, Resistance to mercury and cadmium in *S. aureus* strains of different ecological origin, *Z. Allg. Mikrobiol.*, 517, 1980.

90. **Wolfe, N. L., Zepp, R. G., Gordon, J. A., and Baughman, G. L.**, Chemistry of methylmercurials in aqueous solution, *Chemosphere*, 4, 147, 1973.

91. **Wood, J. M.**, Biological cycles for toxic elements in the environment, *Science*, 183, 1049, 1974.

92. **Wood, J. M., Cheh, A., Dizikes, L. J., Ridley, W. P., Rakow, S., and Lakowicz, J. R.**, Mechanisms for the biomethylation of metals and metalloids, *Fed. Proc.*, 37, 16, 1978.

93. **Wood, J. M., Kennedy, F. S., and Rosen, C. G.**, Synthesis of methylmercury compounds by extracts of a methanogenic bacterium, *Nature (London)*, 220, 173, 1968.

94. **Yamada, M. and Tonomura, K.**, Formation of methylmercury compounds from inorganic mercury by *Clostridium cochlearium*, *J. Ferment. Technol.*, 50, 159, 1972.

95. **Yamada, M. and Tonomura, K.**, Further study of formation of methylmercury from inorganic mercury by *Clostridium cochlearium T-2*, *J. Ferment. Technol.*, 50, 893, 1972.

96. **Yamada, M. and Tonomura, K.**, Microbial methylation of mercury in hydrogen sulfide-evolving environments, *J. Ferment. Technol.*, 50, 901, 1972.

97. **Porter, F.D., Silver, S., and Nakahara, H.**, Selection for mercurial resistance in hospital settings, manuscript in preparation.

98. **Silver, S. and Nakahara, H.**, Bacterial resistance to arsenic compounds, in *Arsenic Syymposium: Production and Use, Biomedical and Environmental Perspectives*, Lederer, W.H., Ed., Van Nostrand Reinhold Co., New York, 1982, in press.

99. **Jackson, W.J. and Summers, A.O.**, Polypeptides encoded by the *mer* operon, *J. Bacteriol*, 149, 479, 1982.

100. **Jackson, W.J. and Summers, A.O.**, The relation of $HgCl_2$-resistance in *Escherichia coli*, Abstracts of the American Society for Microbiology Annual Meeting, p.88, 1982.

101. **NíBhriain, N. and Foster, T.J.**, personal communication.

Chapter 5

RECURRING THEMES AND REPEATED SEQUENCES IN METABOLIC EVOLUTION

L. Nicholas Ornston and Wu-Kuang Yeh

TABLE OF CONTENTS

I. DIVERSITY AND ORDER IN METABOLIC EVOLUTION

Viewed from the perspective of genetic engineering, possibilities for microbial metabolic transformations seem endless. Yet diverse biochemical opportunities preceded molecular biologists and, in response to varied temptations posed during lengthy evolutionary histories, microorganisms seem to have selected restricted metabolic options. Nutritional criteria[1] serve as the basis for taxonomic schemes with high predictive value, and sequence studies[2] suggest that genes do not drift in response to the vicissitudes of life in microbial populations.

Some things microorganisms do, and some they do not. Rules that guide microbial activities may emerge from the study of a single metabolic trait that was successful in diverse protists. Here we describe the beginnings of one such investigation, an attempt to reconstruct fragments of the history of a catabolic pathway that evolved in different microbial genera.

Diversity and order are the recurring themes alluded to in the title of this review. Our history will touch upon diverse induction patterns and varied metabolic mechanisms. Furthermore, we will describe the extensive evolutionary divergence of homologous structural genes. On the other hand, order occurs in the use of specific metabolic pathways by members of some biological groups, the conservation of transcriptional controls within bacterial genera, the maintenance of nearly constant protein structures within bacterial species and, most remarkably, the presence of homologous amino acid sequences in enzymes with different catalytic activities. The repeated sequences form the other half of the title.

II. THE β-KETOADIPATE PATHWAY IN EUCARYOTES AND PROCARYOTES

The subject of our study is the β-ketoadipate pathway,[3,4] a complex set of catabolic reactions employed by protists for the utilization of aromatic and hydroaromatic compounds. As summarized in Figure 1, the central metabolic steps of the pathways begin with either of two diphenols, protocatechuate or catechol. The diphenolic compounds are converted to β-ketoadipate, which is transformed to the tricarboxylic acid cycle intermediates succinate and acetyl CoA. Thus, the β-ketoadipate pathway may be regarded as an addition to the respiratory and assimilatory reactions of the tricarboxylic acid cycle.

Eucaryotes, represented by diverse representatives of the fungi,[5] employ one metabolic variant of the β-ketoadipate pathway and procaryotes, represented by members of Gram-positive[6] and Gram-negative[7] bacterial genera, employ another. As shown in Figure 2, the fungal and bacterial pathways for the utilization of protocatechuate begin with oxygenative cleavage of the diphenol. β-Carboxymuconolactone, the product of ring fission, is converted to β-carboxymuconolactone in fungi[5] and to the isomeric γ-carboxymuconolactone in bacteria[6,7] (Figure 2). A single fungal delactonase[8] hydrolyses and decarboxylates β-carboxymuconolactone to β-ketoadipate (Figure 2). In contrast, two different bacterial enzymes mediate consecutive decarboxylative and hydrolytic reactions as γ-carboxymuconolactone is converted to β-ketoadipate[9] (Figure 2).

The presence of different metabolic mechanisms for dissimilation of protocatechuate suggests that variants of the pathway evolved independently from separate genetic backgrounds in fungi and bacteria. The available data indicate that the respective transformations found in each biological group are mediated by quite different structural gene products. The fungal β-carboxymuconate lactonizing enzyme is a glycoprotein[10,11] formed by association of identical 24,000 dalton subunits into an octamer. Unlike the fungal enzyme, the bacterial β-carboxymuconate lactonizing enzyme does

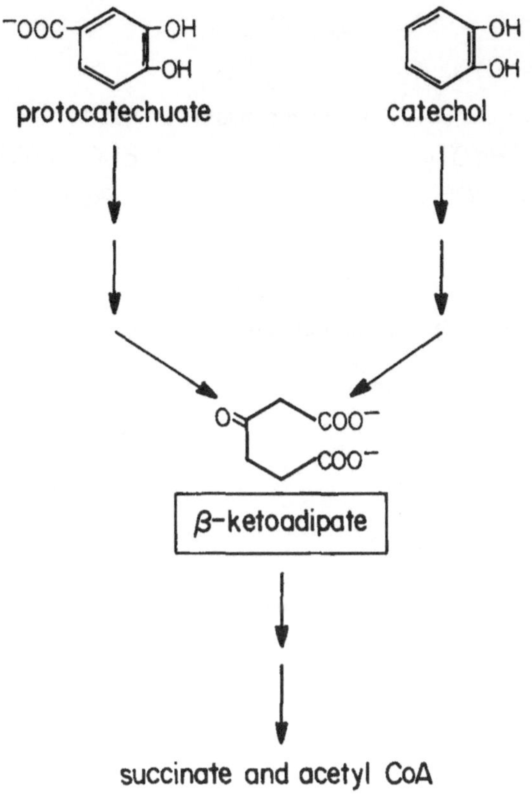

FIGURE 1. Central intermediates in the β-ketoadipate pathway. The pathway has two branches initiating with the diphenols protocatechuate and catechol. Complex growth substrates are converted to one of the diphenolic compounds. The central reactions of the pathway transform protocatechuate and catechol to β-ketoadipate. This metabolite, common to both branches of the pathway, is metabolized to form the tricarboxylic acid cycle intermediates succinate and acetyl CoA.

not appear to contain carbohydrate, and the protein appears to be a hexamer formed by self association of identical 40,000 dalton subunits.[12] Thus the fungal and bacterial enzymes are so dissimilar as to seem the products of convergent evolution, and it would be of interest to know what, if any, chemical and physical properties are shared by enzymes from the two biological groups.

If steps for the utilization of β-carboxymuconate evolved independently in fungi and bacteria, it is possible that protocatechuate 3,4-dioxygenase, the enzyme that gives rise to β-carbomuconate (Figure 2), was derived from separate genetic origins in the two biological groups. Study of the fungal enzyme should provide direct insight into this problem because the physical properties of bacterial protocatechuate 3,4-oxygenase have been explored in detail, and the complete primary structure of one bacterial protocatechuate 3,4-oxygenase has been determined.[13,14]

III. METABOLIC ASPECTS OF THE BACTERIAL β-KETOADIPATE PATHWAY

A. Parallel Metabolic Steps

A striking feature of the bacterial β-ketoadipate pathway is the parallelism of its

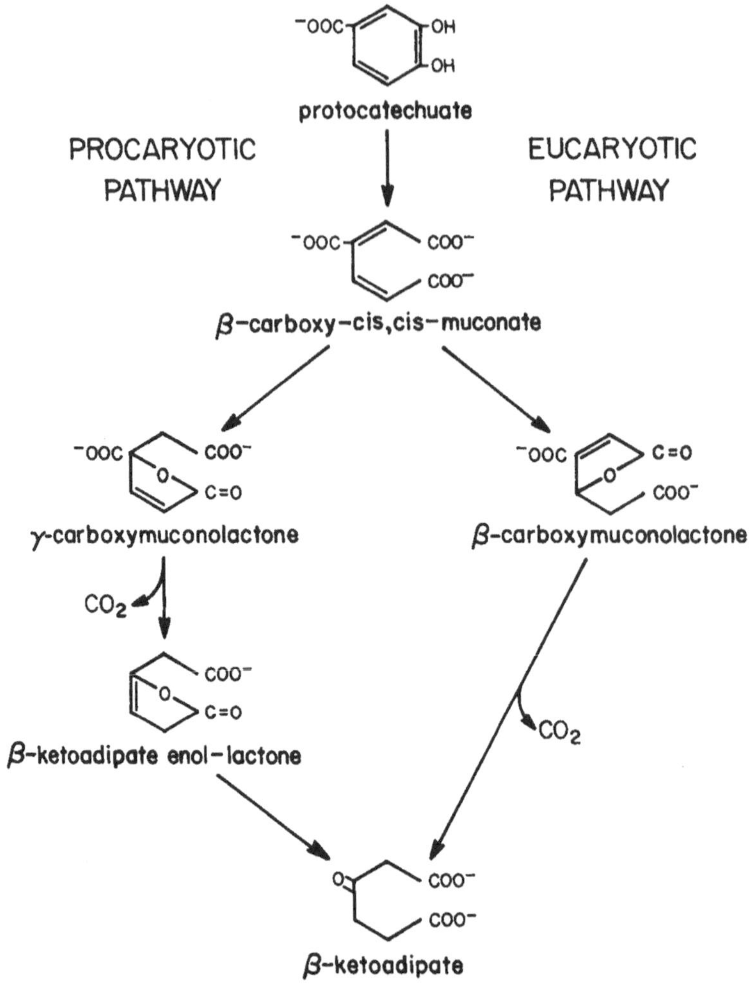

FIGURE 2. Metabolic variants of the protocatechuate branch of the β-ketoadipate pathway. Fungi and bacteria form a protocatechuate 3,4-dioxygenase which gives rise to β-carboxymuconate. This metabolite is converted to β-carboxymuconolactone in eucaryotes and γ-carboxymuconolactone in procaryotes. The two different carboxymuconolactones are metabolized to β-ketoadipate, an intermediate common to fungal and bacterial pathways.

two convergent branches (Figure 3): protocatechuate and catechol differ only by a carboxyl group, and this single chemical difference is maintained in the respective metabolites at each step of the two branches of the pathway. The branches converge metabolically at the level of the β-ketoadipate enol-lactone (Figure 3), but some bacteria give rise to isofunctional enzymes that act on β-ketoadipate and β-ketoadipate enol-lactone. In these organisms one set of enzymes is induced with enzymes of the protocatechuate branch and an isofunctional set of enzymes is induced with enzymes of the catechol branch. The analogy or identity of enzymes mediating parallel steps in the β-ketoadipate pathway raises the possibility that ancient gene duplication may have given rise to ancestral genes for the two branches of the pathway. We shall present evidence suggesting that this is indeed the case in Sections V and VI.

B. Bacteria Possessing Fragments of the β-Ketoadipate Pathway

The full suite of enzymes for both branches of the β-ketoadipate pathway is found

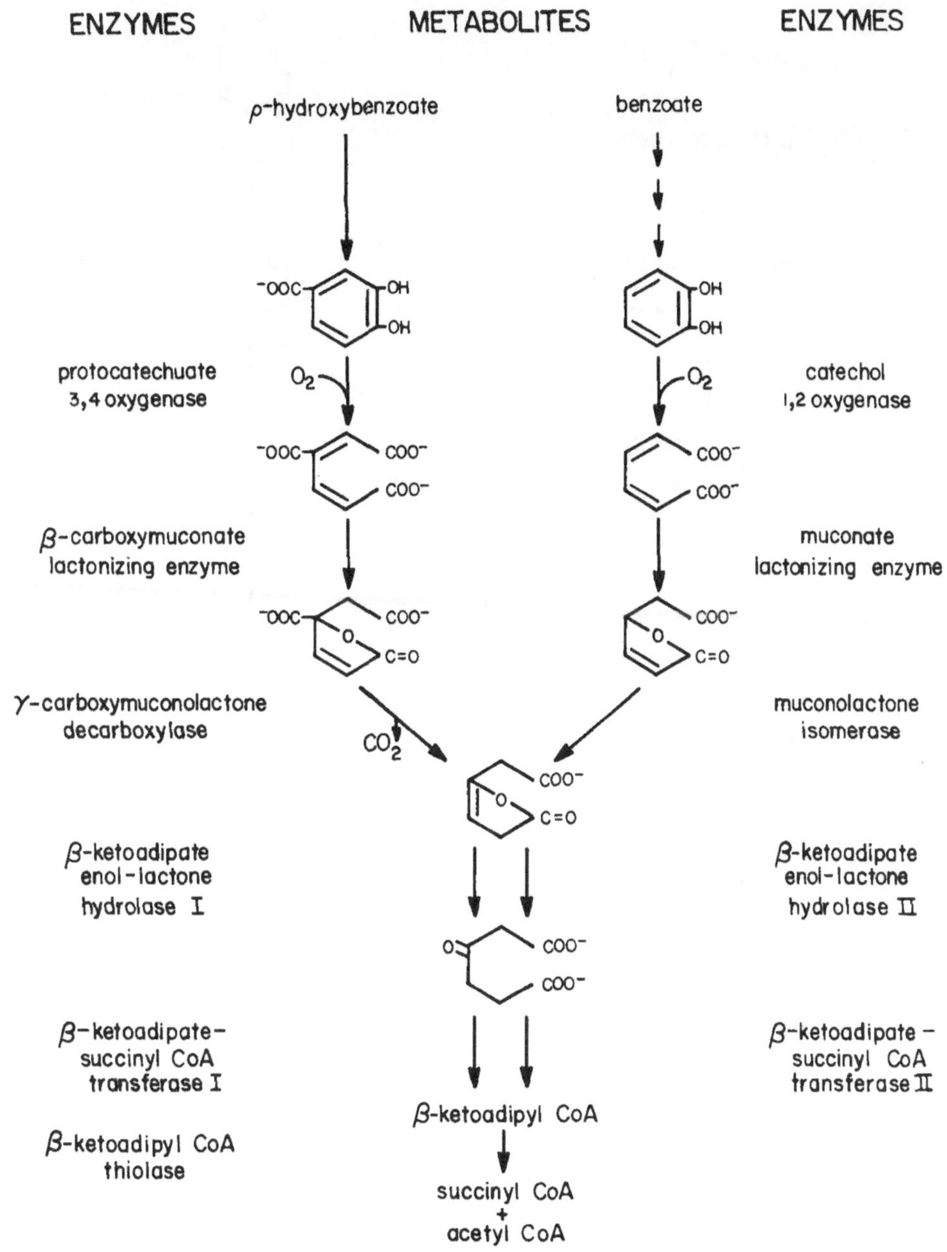

FIGURE 3. Central steps of the β-ketoadipate pathway in bacteria. Analogous metabolic transformations convert protocatechuate and catechol to β-ketoadipate enol-lactone. Induction patterns used by members of some bacterial genera require the synthesis of isofunctional enzymes for the steps that follow metabolic convergence of the two branches of the pathway at β-ketoadipate enol-lactone.

in representatives of many bacterial genera, but other organisms possess enzymes for only portions of the pathway. For example, members of *Azotobacter* utilize protocatechuate via the β-ketoadipate pathway and metabolize catechol by a completely independent metabolic route (Figure 4).[15,16] *Pseudomonas acidovorans* and *P. testosteroni* (organisms classified in the Acidovorans *Pseudomonas* group) do not form the intradiol cleaving dioxygenases (ortho dioxygenases) associated with the β-ketoadipate pathway.[1] As shown in Figure 5, these organisms subject protocatechuate and catechol

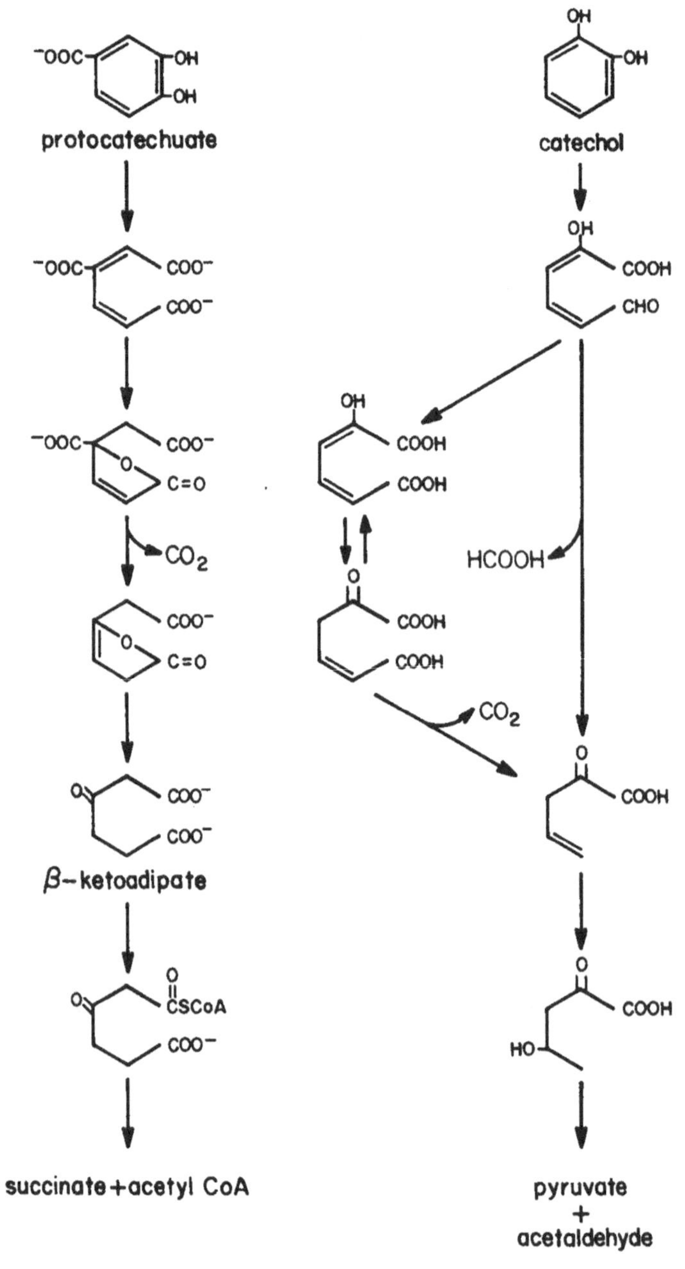

FIGURE 4. Pathways for the dissimilation of protocatechuate and catechol in *Azotobacter*. Representatives of this genus utilize protocatechuate via the β-ketoadipate pathway and catechol via a different series of reactions which give rise to pyruvate and acetaldehyde.

to extradiol oxygenative cleavage (mediated by meta oxygenases) and metabolize the products of ring fission to pyruvate and acetaldehyde.[17] Members of the Acidovorans group grow well at the expense of the muconic acids formed by the ortho cleavage of protocatechuate and catechol (Figure 5), and thus these bacteria form all of the enzymes of the β-ketoadipate pathway except the ortho dioxygenases.[18,19] The muconic acids do not readily permeate the membrane of most bacteria that use the β-ketoadipate pathway;[3,20] in these organisms the muconic acids serve solely as the intermediates

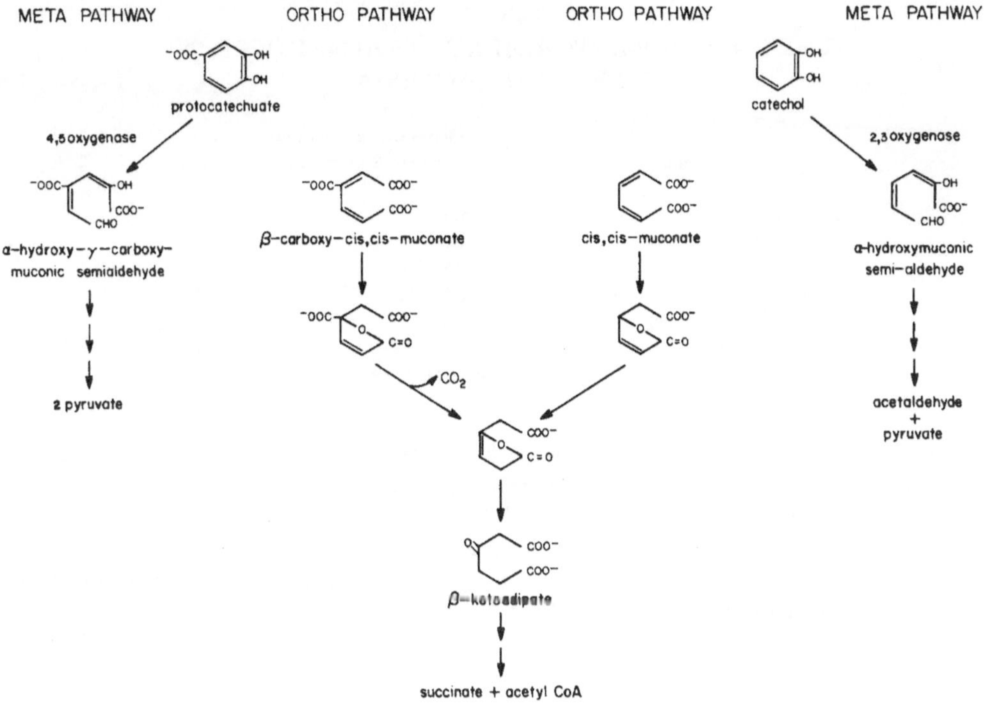

FIGURE 5. The β-ketoadipate pathway is not used for the dissimilation of aromatic acids in members of the Acidovorans · *Pseudomonas* group. In these organisms metabolism of protocatechuate and catechol is initiated by extradiol oxygenative cleavage giving rise to muconic semialdehydes, which are converted to pyruvate and acetaldehyde. The organisms use the β-ketoadipate pathway to grow at the expense of the muconic acids which evidently serve as growth substrates in the niche of this bacterial group.

in the dissimilation of aromatic and hydroaromatic compounds. Hence the niche of *Pseudomonas* species in the Acidovorans group is distinguished by the presence of muconic acids as growth substrates. In this case comparison of metabolic pathway helps to define ecological factors that shape the properties of a specific group of bacteria.

C. Comparison of Induction Patterns

Enzymes of the β-ketoadipate pathway are inducible.[3,4] Within any organism a small set of metabolites, known as inducers, elicits the synthesis of enzymes. Comparative studies of the β-ketoadipate pathway have shown that a single inducer may cause formation of only one enzyme[21] or may lead to the synthesis of as many as nine enzymes.[22] Members of a single biological group normally share a sharply defined induction pattern, but the full range of induction patterns varies widely. As an example, we shall consider enzymes induced by β-ketoadipate in different biological groups.

Structural genes for the β-ketoadipate pathway are represented in Figure 6, and the structural genes that are expressed in response to β-ketoadipate are summarized in Table 1. It can be seen that the capacity of β-ketoadipate to act as an inducer varied widely. In *Nocardia* the compound induces all of the enzymes of the protocatechuate pathway and two enzymes of the catechol pathway; in *Acinetobacter* β-ketoadipate induces none of the enzymes; members of other biological groups fall into intermediate classes (Table 1).

Transcriptional controls that underlie the induction patterns of bacteria presumably are highly evolved traits, and highly evolved traits tend to be conserved. For example,

Table 1
THE ROLE OF β-KETOADIPATE AS AN INDUCER IN DIFFERENT BACTERIA

Genus	Genes expressed in response to β-ketoadipate	Ref.
Acinetobacter	None	23
Alcaligenes	*pcaE*	24
Azotobacter	*pcaBCDE*	25
Pseudomonas (Fluorescent and Cepacia groups)	*pcaBCDE*	20
P. testosteroni	*catC; cat DE* or *pcaDE*	19
Nocardia	*catCD; pcaABCDE*	6

metabolic variants of the β-ketoadipate pathway (Section II), one successful in eucaryotes and the other successful in procaryotes, appear to have risen from separate genetic origins and to have been conserved in each biological group. We now consider sets of regulatory genes which, as indicated by induction patterns, are conserved within different procaryotic groups. Did the regulatory genes arise from separate genetic origins? No direct evidence can be brought to bear on this question, but comparative studies demonstrate the common ancestry of structural genes for the pathway in different bacterial genera. For example, carboxymuconolactone decarboxylase, the product of the *pcaC* gene, is induced by protocatechuate in *Acinetobacter calcoaceticus* and by β-ketoadipate in *Pseudomonas putida*.[3,4] As described in the following section, the isofunctional decarboxylases are unmistakably homologous. This, and additional evidence (Section IV), force the conclusion that the β-ketoadipate pathway was derived from a single set of ancestral genes in those procaryotic genera that have been examined. It may well be that the organisms possess homologous regulatory genes and the different induction patterns were achieved by transposition of structural and regulatory genes. Similar events have been proposed by Crawford[26] to account for diverse transcriptional controls exerted over the enzymes that mediate tryptophan biosynthesis.

Transposition mutations are not unusual, and it is reasonable to ask why they do not continue to provide a range of induction patterns by ongoing mutations within any single taxon. The answer to this question must lie outside the organisms (the natural selection of distinctive induction patterns in different niches), within the organisms (organization of the regulatory regions so that they are relatively immutable), or perhaps both places.

IV. ISOFUNCTIONAL ENZYMES FROM DIFFERENT BACTERIAL GENERA ARE EVOLUTIONARILY HOMOLOGOUS

A. Protocatechuate 3,4-Dioxygenase

Protocatechuate 3,4-dioxygenase (EC 1.13.1.3) is the most thoroughly studied enzyme of the β-ketoadipate pathway. Bacterial protocatechuate 3,4-dioxygenase is formed by self association of equal amounts of nonidentical α and β subunits each of which has a molecular weight of approximately 23,000 daltons.[27-31] The complete amino acid sequences of both subunits of the *P. aeruginosa* oxygenases are known,[13,14] and the two primary structures appear to be quite different. Protocatechuate 3,4-dioxygenase contains a single atom of iron per αβ protomer, and Mossbauer spectroscopy has been used to elucidate the mechanism of the enzyme.[32] The mechanism of the oxygenase is discussed in Chapter 3 of this volume.

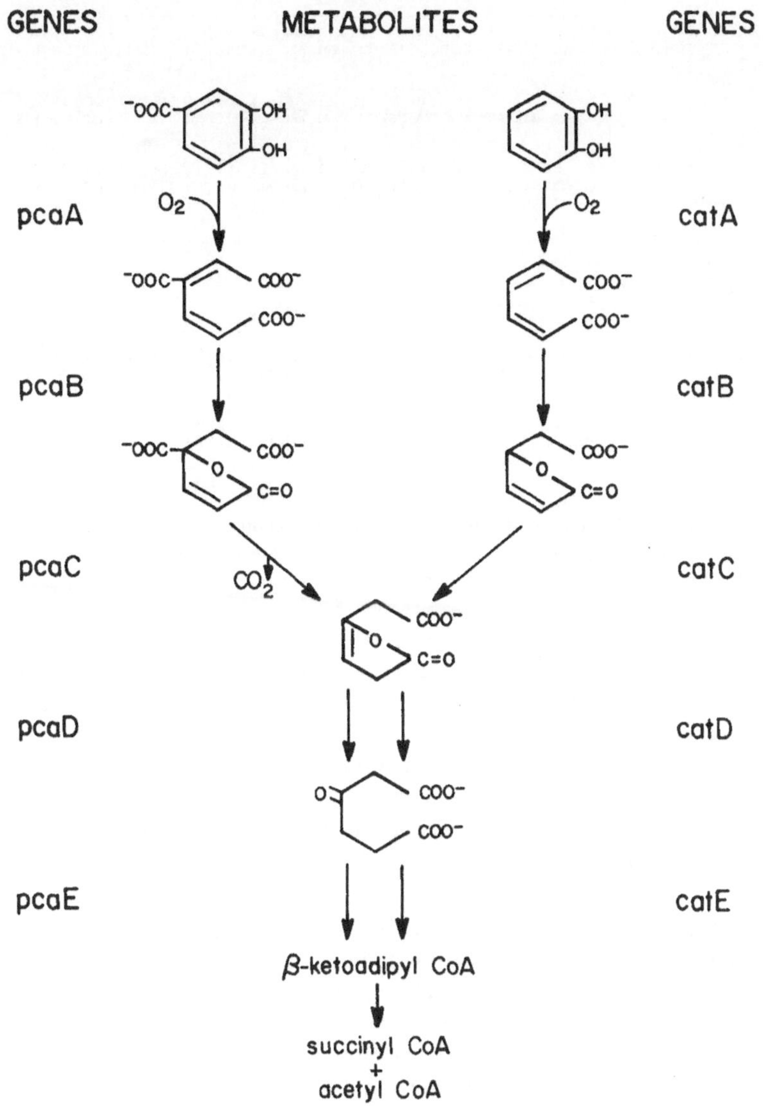

FIGURE 6. Structural genes for steps in the β-ketoadipate pathway. The designations depicted here may be used to identify the genes that are expressed in response to the inducer β-ketoadipate (Table 1).

The oligomeric structure of bacterial protocatechuate oxygenases varies: the enzyme has been obtained from different biological sources, and the number of $\alpha\beta$ protomers per oligomer has been reported to be 4,[30] 6,[29] 8,[28] and 10.[31] Despite their varied oligomeric sizes, protocatechuate 3,4-dioxygenases from *Acinetobacter, Azotobacter* and *Pseudomonas* cross-react immunologically, and this evidence indicates that the proteins all share a common ancestral gene.[31]

B. Carboxymuconolactone Decarboxylase (EC 4.1.1.44)

Carboxymuconolactone decarboxylase is the enzyme from the β-ketoadipate pathway for which the most comparative amino acid sequence data are available. The decarboxylase from *Acinetobacter, Azotobacter* and *Pseudomonas* exists as a hexamer composed of identical 13,000-dalton subunits. As shown in Figure 7, the NH_2-terminal

	1	2	3	4	5	6	7	8	9	10	11	12	13	14	15	16	17	18	19	20	21	22	23	24	25
Acinetobacter Decarboxylase	MET	ASN	ASP	GLU	GLN	ARG	TYR	LYS	GLN	GLY	LEU	GLU	VAL	ARG	THR	GLU	VAL	LEU	GLY	GLU	LYS	HIS	VAL	ASN	ARG
Azotobacter Decarboxylase	MET	ASP	GLU	LYS	GLU	ARG	TYR	ASP	ALA	GLY	MET	GLN	VAL	ARG	ARG	ALA	VAL	LEU	GLY	ASP	ALA	HIS	VAL	ASP	ARG
Pseudomonas Decarboxylase	MET	ASP	GLU	LYS	GLN	ARG	TYR	ASP	ALA	GLY	MET	GLN	VAL	ARG	ARG	ALA	VAL	LEU	GLY	ASP	ALA	HIS	VAL	ASP	ARG

	26	27	28	29	30	31	32	33	34	35	36	37	38	39	40	41	42	43	44	45	46	47	48	49
Acinetobacter Decarboxylase	SER	LEU	GLU	ASN	LEU	ASP	ASP	PHE	ARG (?)	GLN	ASP													
Azotobacter Decarboxylase	CYS (?)	LEU	LYS	ASN	LEU	THR	PRO	PHE	ASN	GLU	GLU	PHE	GLN	GLU	MET	ILE	THR	ARG	HIS	ALA (?)	CYS (?)	GLY	ASP	ILE
Pseudomonas Decarboxylase	CYS (?)	LEU	GLU	LYS	LEU	ASN	ASP	PHE	ASN	GLY	GLU	PHE	GLN	GLU	MET	ILE	THR	ARG	HIS	ALA	CYS (?)	GLY	ASP	ILE

FIGURE 7. NH_2-Terminal amino acid sequences of carboxymuconolactone decarboxylases from *Acinetobacter*, *Azotobacter* and *Pseudomonas*. Numerals above amino acid residues indicate their positions in the primary structures of the proteins. Identical residues are enclosed in boxes.

amino acid sequences of the *Azotobacter* and *Pseudomonas* decarboxylases are closely homologous, and these two sequences are more distantly related to the NH_2-terminal amino acid sequence of the *Acinetobacter* decarboxylase. *Azotobacter*[25] and members of the fluorescent *Pseudomonas* group[20] share similar induction patterns in which β-ketoadipate acts as an inducer of the decarboxylase (Table 1). The relatively dissimilar decarboxylase of *Acinetobacter* is induced by protocatechuate.[23] Thus it appears that the most widely divergent decarboxylase genes have been placed under control exercised by different regulatory genes.

C. Muconolactone Isomerase (EC 5.3.3.4)

Muconolactone isomerase from *Acinetobacter* and *Pseudomonas* appears to be either a decamer or a dodecamer in each case formed by self association of identical 11,000-dalton subunits.[33,36] As shown in Figure 8, the NH_2-terminal amino acid sequences of the isomerases[37] are clearly homologous; more than half of the aligned residues are identical, and the compared sequences correspond to more than half of the primary structures of the proteins.

D. Immunological Comparisons

For the most part immunological comparison of enzymes of the β-ketoadipate pathway has shown serological relatedness of isofunctional proteins formed by different representatives of a single bacterial genus.[38-41] With few exceptions, immunological techniques fail to show evolutionary relationships among isofunctional enzymes formed by members of different genera. Interesting exceptions to the rule are members of *Azotobacter* which form a protocatechuate 3,4-dioxygenase that cross-reacts strongly with the isofunctional enzyme from fluorescent *Pseudomonas* and less strongly with *Acinetobacter* protocatechuate 3,4-dioxygenase.[31] Other exceptions are strains classified in the Acidovorans *Pseudomonas* group. These bacteria do not share the induction pattern characteristic of fluorescent *Pseudomonas species,*[19] and enzymes mediating reactions of the β-ketoadipate pathway in the Acidovorans group do not exhibit immunological cross-reaction with isofunctional proteins from fluorescent pseudomonads.[41] Representatives of *Azotobacter* have only one branch of the β-ketoadipate pathway[15] (Section III.B), and enzymes of this branch are induced by a

	1	2	3	4	5	6	7	8	9	10	11	12	13	14	15	16	17	18	19	20	21	22	23	24	25
Acinetobacter Isomerase	MET	LEU	PHE	GLN	VAL	ARG	MET	ASP	VAL	HIS	LEU	PRO	VAL	SER	MET	PRO	THR	ASP	GLN	ALA	ASN	GLN	ILE	LYS	SER
Pseudomonas Isomerase	MET	LEU	PHE	HIS	VAL	LYS	MET	THR	VAL	LYS	LEU	PRO	VAL	ASP	MET	ASP	PRO	ALA	LYS	ALA	THR	GLN	LEU	LYS	ALA

	26	27	28	29	30	31	32	33	34	35	36	37	38	39	40	41	42	43	44	45	46	47	48	49	50
Acinetobacter Isomerase	VAL	GLU	LYS	ALA	TYR	SER	GLN	GLU	LEU	GLN	ARG	GLN	GLY	LYS	TRP	ARG	HIS	ILE	TRP	ARG	ILE	THR	GLY	GLN	TYR
Pseudomonas Isomerase	ASP	GLU	LYS	GLU	LEU	ALA	GLN	ARG	LEU	GLN	ARG	GLU	GLY	THR	TRP	ARG	HIS	LEU	TRP	ARG	ILE	ALA	GLY	HIS	TYR

FIGURE 8. NH_2-terminal amino acid sequences of muconolactone isomerases from *Acinetobacter* and *Pseudomonas.* Numerals above amino acid residues indicate their positions in the primary structures of the proteins. Identical residues are enclosed in boxes.

mechanism indistinguishable from that found in fluorescent pseudomonads.[25] Thus with respect to the β-ketoadipate pathway, the structural and regulatory genes of *Azotobacter* resemble those of fluorescent *Pseudomonas* species more closely than genes from the latter organisms resemble the corresponding genes in *Pseudomonas* species of the Acidovorans group.

Intrageneric immunological comparisons of enzymes of the β-ketoadipate pathways have been conducted primarily with antisera prepared against reference proteins from fluorescent *Pseudomonas* species. The most remarkable trait to emerge from these studies is the conservation of sets of closely related antigenic determinants among isofunctional enzymes formed by members of a single species or biotype.[38-41] Observed against a background of wide variation in the serological properties of isofunctional enzymes within the genus *Pseudomonas,* the observed conservation of antigenic determinants within biological subgroups rules out the possibility that genes for the β-ketoadipate pathway continue to be exchanged among different *Pseudomonas* species.[41] Evidently exchange of genes for the pathway among *Pseudomonas* species stopped some time ago, and since then the genes have stabilized in a form consonant with the niche and genome of their hosts. It is noteworthy that, throughout the multiple waxings and wanings of bacterial populations, the genes have shown little tendency to drift.

V. CO-EVOLVING GENES FOR ISOFUNCTIONAL ENZYMES HAVE DIVERGED WIDELY FROM A COMMON EVOLUTIONARY ORIGIN

Insofar as we know, enzymes of the β-ketoadipate pathway are not regulated at the level of their activity. Therefore selective pressures do not appear to have been exerted for divergence of isofunctional enzymes to accommodate different allosteric ligands. In view of this evidence, described above, that bacterial genes tend not to drift, it is remarkable that genes for the isofunctional β-ketoadipate enol-lactone hydrolases of *Acinetobacter* have diverged to a considerable extent. The isofunctional proteins are oligomers of comparable size formed by self association of identical subunits of about 12,000 daltons.[42] Yet, as shown in Figure 9, the NH_2-terminal amino acid sequences of the enzymes contain divergent residues in about 80% of the 30 positions analyzed.[43] Indeed convincing evidence that the two enol-lactone hydrolases share a common ancestor comes only when the NH_2-terminal amino acid sequence of the single β-ketoadipate enol-lactone hydrolase formed by *Pseudomonas*[44] is included in the comparison. Examination of Figure 10 shows that many of the residues in one or both of the *Acinetobacter* hydrolase sequences are identical to residues occupying the corresponding positions in the *Pseudomonas* hydrolase sequence. A reasonable interpretation of

Acinetobacter Hydrolase I	1 CYS	2 GLU	3 ILE	4 HIS	5 SER	6 ILE	7 MET	8 ILE	9 THR	10 ASN	11 ARG	12 GLN	13 GLY	14 LYS	15 THR
Acinetobacter Hydrolase II	1 PRO	2 VAL	3 PHE	4 HIS	5 PHE	6 LYS	7 ASP	8 THR	9 LEU	10 THR	11 ALA	12 GLN	13 ASP	14 VAL	15 ALA

Acinetobacter Hydrolase I	16 LEU	17 SER	18 VAL	19 GLX	20 ILE	21 ASX	22 TYR	23 PRO	24 GLX	25 ASX	26 PRO (?)	27 PRO	28 ALA (?)	29 ILE	30 VAL
Acinetobacter Hydrolase II	16 LEU	17 ASP	18 TYR	19 ALA	20 THR	21 PHE	22 GLY	23 GLN	24 ALA	25 ASP	26 ARG	27 PRO	28 ALA	29 LEU	30 ILE

FIGURE 9. NH_2-Terminal amino acid sequences of *Acinetobacter* enol-lactone hydrolases I and II. Numerals above amino acid residues indicate their positions in the primary structures of the proteins. Identical residues are enclosed in boxes. It is evident that the two sequences have diverged widely.

Acinetobacter Hydrolase I	1 CYS	2 GLU	3 ILE	4 HIS	5 SER	6 ILE	7 MET	8 ILE	9 THR	10 ASN	11 ARG	12 GLN	13 GLY	14 LYS	15 THR
Pseudomonas Hydrolase							1 ALA	2 HIS	3 LEU	4 GLN	5 LEU	6 ALA	7 ASP	8 GLY	9 VAL
Acinetobacter Hydrolase II	1 PRO	2 VAL	3 PHE	4 HIS	5 PHE	6 LYS	7 ASP	8 THR	9 LEU	10 THR	11 ALA	12 GLN	13 ASP	14 VAL	15 ALA

Acinetobacter Hydrolase I	16 LEU	17 SER	18 VAL	19 GLX	20 ILE	21 ASX	22 TYR	23 PRO	24 GLX	25 ASX	26 PRO (?)	27 PRO	28 ALA (?)	29 ILE	30 VAL
Pseudomonas Hydrolase	10 LEU	11 ASN	12 TYR	13 GLN	14 ILE	15 ASP	16 GLY	17 PRO	18 GLU	19 ASN	20 ALA	21 PRO	22 VAL	23 LEU	24 VAL
Acinetobacter Hydrolase II	16 LEU	17 ASP	18 TYR	19 ALA	20 THR	21 PHE	22 GLY	23 GLN	24 ALA	25 ASP	26 ARG	27 PRO	28 ALA	29 LEU	30 ILE

FIGURE 10. NH_2-Terminal amino acid sequences of β-ketoadipate enol-lactone hydrolases. Inclusion of the *Pseudomonas* hydrolase sequence (depicted here) in the comparison of the *Acinetobacter* enol-lactone hydrolases (depicted in Figure 9) demonstrates the common ancestry of all three sequences. Numerals above amino acid residues indicate their positions in the primary structures of the proteins. Identical residues are enclosed in boxes.

these observations is that an ancestral β-ketoadipate enol-lactone hydrolase gene, containing sequence segments still represented in the *Pseudomonas* hydrolase gene, duplicated to form the respective ancestral genes for *Acinetobacter* enol-lactone hydrolases I and II.

The duplicated enol-lactone hydrolase genes diverged widely as they co-evolved in *Acinetobacter*, and it is instructive to consider selective forces that favor the extensive divergence of homologous genes as they co-evolve within a single cell line.[45] One of the selective forces is recombination: in order to be genetically stable in the presence of recombination enzymes, homologous genes must differ in their DNA sequences. As long as the different nucleotide sequences do not lead to amino acid replacements causing loss of function, they will not be counterselected. As exemplified by *Acinetobacter* enol-lactone hydrolases I and II, functional proteins may vary widely in amino

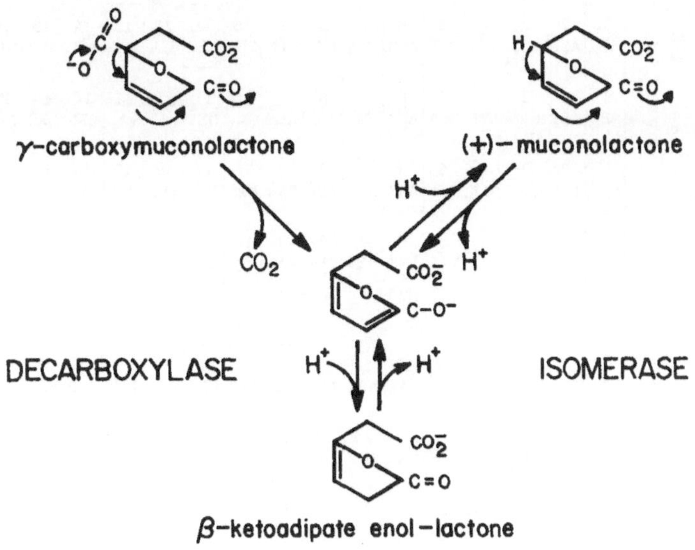

FIGURE 11. Chemically analogous intralactonic rearrangements mediated by carboxymuconolactone decarboxylase and muconolactone isomerase.

acid sequence. A second selective force favoring the evolutionary divergence of homologous proteins is the accurate self association of protomers into oligomers. Conservation of the full set of amino acid residues at protein-protein interfaces within oligomers could cause homologous protomers to associate into hybrid oligomers. Such proteins are not observed among enzymes of the β-ketoadipate pathway, all of which are oligomeric.

It is likely that selective forces favoring divergence away from homologous sequences are less discriminating than forces favoring acquisition of a specific binding site or catalytic site in a protein. The latter forces may rightly be considered sources of genetic order because they favor the convergence of analogous sequences or the conservation of established sequences. In contrast, selection for the substitution of nucleotide or amino acid sequences into homologous genes may act as a source of genetic diversity.

VI. CO-EVOLVING GENES FOR ENZYMES MEDIATING CHEMICALLY ANALOGOUS REACTIONS HAVE DIVERGED WIDELY FROM A COMMON EVOLUTIONARY ORIGIN

As shown in Figure 11, carboxymuconolactone decarboxylase and muconolactone isomerase mediate chemically analogous intralactonic rearrangements. The former enzyme, unlike the latter, couples the migration of a double bond to a decarboxylation. The chemical analogy of the two-type reactions suggest that the decarboxylase and the isomerase might share a common ancestral gene, and this inference is supported by comparison of the NH_2-terminal amino acid sequences of the enzymes.[34] As shown in Figure 12, the available sequence data allow a number of pairwise comparisons to be made. On average, the level of sequence identity in the decarboxylase-isomerase comparisons (Figure 12) is about 20%, roughly the same value observed in the comparison of the NH_2-terminal amino acid sequences of *Acinetobacter* enol-lactone hydrolases I and II (Figure 9). Comparison of *Pseudomonas* decarboxylase and *Acinetobacter* isomerase (Figure 12) reveals sequence identity of 28%, far above the level that would be observed in the comparison of unrelated sequences. In the *Pseudomonas* decarboxyl-

Acinetobacter Decarboxylase	1 MET	2 ASN	3 ASP	4 GLU	5 GLN	6 ARG	7 TYR	8 LYS	9 GLN	10 GLY	11 LEU	12 GLU	13 VAL	14 ARG	15 THR	16 GLU	17 VAL
Pseudomonas Decarboxylase	1 MET	2 ASP	3 GLU	4 LYS	5 GLN	6 ARG	7 TYR	8 ASP	9 ALA	10 GLY	11 MET	12 GLN	13 VAL	14 ARG	15 ARG	16 ALA	17 VAL
Acinetobacter Isomerase	1 MET	2 LEU	3 PHE	4 GLN	5 VAL	6 ARG	7 MET	8 ASP	9 VAL	10 HIS	11 LEU	12 PRO	13 VAL	14 SER	15 MET	-Δ-	16 PRO
Pseudomonas Isomerase	1 MET	2 LEU	3 PHE	4 HIS	5 VAL	6 LYS	7 MET	8 THR	9 VAL	10 LYS	11 LEU	12 PRO	13 VAL	14 ASP	15 MET	16 ASP	17 PRO

Acinetobacter Decarboxylase	18 LEU	19 GLY	20 GLU	21 LYS	22 HIS	23 VAL	24 ASN	25 ARG	26 SER	27 LEU	28 GLU	29 ASN	30 LEU	31 ASP	32 ASP	33 PHE	34 ARG (?)
Pseudomonas Decarboxylase	18 LEU	19 GLY	20 ASP	21 ALA	22 HIS	23 VAL	24 ASP	25 ARG	26 CYS (?)	27 LEU	28 GLU	29 LYS	30 LEU	31 ASN	32 ASP	33 PHE	34 ASN
Acinetobacter Isomerase	17 THR	18 ASP	19 GLN	20 ALA	21 ASN	22 GLN	23 ILE	24 LYS	25 SER	26 VAL	27 GLU	28 LYS	-Δ-	-Δ-	29 ALA	30 TYR	31 SER
Pseudomonas Isomerase	-Δ-	18 ALA	19 LYS	20 ALA	21 THR	22 GLN	23 LEU	24 LYS	25 ALA	26 ASP	27 GLU	28 LYS	-Δ-	-Δ-	29 GLU	30 LEU	31 ALA

Acinetobacter Decarboxylase	35 GLN	36 ASP													
Pseudomonas Decarboxylase	35 GLY	36 GLU	37 PHE	38 GLN	39 GLU	40 MET	41 ILE	42 THR	-Δ-	43 ARG	44 HIS	45 ALA	46 CYS (?)	47 GLY	48 ASP
Acinetobacter Isomerase	32 GLN	33 GLU	34 LEU	35 GLN	36 ARG	37 GLN	38 GLY	39 LYS	40 TRP	41 ARG	42 HIS	43 ILE	44 TRP	45 ARG	46 ILE
Pseudomonas Isomerase	32 GLN	33 ARG	34 LEU	35 GLN	36 ARG	37 GLU	38 GLY	39 THR	40 TRP	41 ARG	42 HIS	43 LEU	44 TRP	45 ARG	46 ILE

FIGURE 12. NH_2-Terminal amino acid sequences of carboxymuconolactone decarboxylases and muconolactone isomerases from *Acinetobacter* and *Pseudomonas.* Several gaps have been introduced to allow optimal alignment of identical residues in the decarboxylase-isomerase comparisons; the identical residues are enclosed in boxes. Numerals above amino acid residues indicate their positions in the primary structures of the proteins.

ase-*Acinetobacter* isomerase comparison, six identical residues are aligned by introduction of a single gap into the region between residues 21 and 40 of the *Pseudomonas* decarboxylase sequence. The probability that this level of sequence identity could be due to chance alone is less than 0.025;[46] the probability that the sequence identity observed in all of the decarboxylase-isomerase comparisons is due to chance alone is vanishingly small. In principle, the similarity of the decarboxylase and isomerase sequences might be attributed to either convergent or divergent evolution. The former interpretation, that the similar sequences are due to convergent evolution of similar catalytic sites in the decarboxylases and the isomerases, is undermined by observation of chemically dissimilar substitutions in regions where pairwise decarboxylase-isomerase comparisons show sequence identity. For example, the aspartyl residue shared by *Pseudomonas* decarboxylase and *Acinetobacter* isomerase at position 8 is substituted by the chemically dissimilar lysine in position 8 of the *Acinetobacter* decarboxylase sequence (Figure 12). Chemically dissimilar substitutions also are observed at sites of sequence identity corresponding to positions 21, 28, 32, 33, and 39 of the *Acinetobac-*

Pseudomonas Isomerase	1	2	3	4	5	6	7	8	9	10	11	12	13	14	15
	MET	LEU	PHE	HIS	VAL	LYS	MET	THR	VAL	LYS	LEU	PRO	VAL	ASP	MET
Acinetobacter Hydrolase II	1	2	3	4	5	6	7	8	9	10		11	12	13	14
	PRO	VAL	PHE	HIS	PHE	LYS	ASP	THR	LEU	THR	-Δ-	ALA	GLN	ASP	VAL

Pseudomonas Isomerase	16	17	18	19	20	21	22	23	24	25	26	27	28	29	30
	ASP	PRO	ALA	LYS	ALA	THR	GLN	LEU	LYS	ALA	ASP	GLU	LYS	GLU	LEU
Acinetobacter Hydrolase II	15	16	17	18	19	20	21	22	23	24	25	26	27	28	29
	ALA	LEU	ASP	TYR	ALA	THR	PHE	GLY	GLN	ALA	ASP	ARG	PRO	ALA	LEU

FIGURE 13. NH_2-Terminal amino acid sequences of *Pseudomonas* muconolactone isomerase and *Acinetobacter* enol-lactone hydrolase II. A single gap has been introduced into the hydrolase sequence to allow optimal alignment of identical residues; these are enclosed in boxes. Numerals above amino acid residues indicate their positions in the primary structures of the proteins.

ter isomerase sequence. Thus the data favor the conclusion that the decarboxylase and isomerase sequence possess similarities indicative of a common evolutionary origin; the sequences appear to have diverged widely from a single ancestor.

To some extent sequence divergence appears to be reflected in the different oligomeric structures of the decarboxylases and isomerases. The decarboxylases appear to be hexameric,[33,34] whereas the isomerases exhibit the properties of either decamers formed by association of pentamers or dodecamers formed by association of hexamers.[33,47]

VII. SIMILAR AMINO ACID SEQUENCES OCCUR IN ENZYMES WITH DIFFERENT CATALYTIC ACTIVITIES

Muconolactone isomerase and β-ketoadipate enol-lactone hydrolase mediate consecutive metabolic reactions (Figure 3). The subunit molecular weights of the enzymes are between 10,000 and 13,000 daltons. As shown in Figure 13, the NH_2-terminal amino acid sequences of *Pseudomonas* muconolactone isomerase and *Acinetobacter* enol-lactone hydrolase II are evolutionarily homologous.[37] One, not necessarily accurate, interpretation of these findings is that duplication of a single ancestral gene gave rise to the ancestral genes for muconolactone isomerase and enol-lactone hydrolase. A problem with this interpretation is that the enol-lactone hydrolases have diverged considerably from their common ancestry (Figures 9 and 10), and the isomerases have diverged to a great extent from the ancestral sequence that they share with carboxymuconolactone decarboxylases (Figure 12). It is difficult to imagine how the similar amino acid sequences found in the NH_2-terminal amino acid sequences of *Pseudomonas* muconolactone isomerase and *Acinetobacter* enol-lactone hydrolase II (Figure 13) could have been conserved against a background of such extreme genetic divergence.

Another interpretation of the *Pseudomonas* isomerase-*Acinetobacter* enol-lactone hydrolase II sequence similarity is that DNA sequences from an ancestral isomerase gene may have been substituted into the ancestral *Acinetobacter* hydrolase II gene as it diverged from the ancestral hydrolase gene.[48] The notion that sequences were transferred from an ancestral isomerase gene to diverging hydrolase genes is fortified by observation of similar sequences in the NH_2-terminal segments of *Acinetobacter* muconolactone isomerase and *Acinetobacter* enol-lactone hydrolase I[43] (Figure 14). Aside

Acinetobacter Isomerase	9	10	11	12	13	14	15	16	17	18	19	20	21	22	23
	VAL	HIS	LEU	PRO	VAL	SER	MET	PRO	THR	ASP	GLN	ALA	ASN	GLN	ILE
Acinetobacter Hydrolase I	1	2	3	4	5	6	7	8	9	10	11	12	13	14	15
	CYS	GLU	ILE	HIS	SER	ILE	MET	ILE	THR	ASN	ARG	GLN	GLY	LYS	THR
Acinetobacter Isomerase	26	27	28	29	30	31	32	33	34	35	36	37	38	39	40
	VAL	GLU	LYS	ALA	TYR	SER	GLN	GLU	LEU	GLN	ARG	GLN	GLY	LYS	TRP

Acinetobacter Isomerase	24	25	26	27	28	29	30	31	32	33	34	35	36	37	38
	LYS	SER	VAL	GLU	LYS	ALA	TYR	SER	GLN	GLU	LEU	GLN	ARG	GLN	GLY
Acinetobacter Hydrolase I	16	17	18	19	20	21	22	23	24	25	26	27	28	29	30
	LEU	SER	VAL	GLX	ILE	ASX	TYR	PRO	GLX	ASX	PRO	PRO	ALA	ILE	VAL
Acinetobacter Isomerase	41	42	43	44	45	46	47	48	49	50					
	ARG	HIS	ILE	TRP	ARG	ILE	THR	GLY	GLN	TYR					

FIGURE 14. NH_2-Terminal amino acid sequences of *Acinetobacter* muconolactone isomerase and *Acinetobacter* enol-lactone hydrolase I. Numerals above amino acid residues indicate their positions in the primary structures of the proteins. Residues 9 and 26 in the *Acinetobacter* isomerase sequence have been aligned with Residue 1 of the *Acinetobacter* hydrolase I sequence to achieve optimal alignment of identical residues; these are enclosed in boxes.

from their occurrence, the remarkable feature of these sequence similarities is that their observation requires shifts in the relative alignment of isomerase and hydrolase sequences: in Figure 13 Residue 1 of the isomerase is aligned with Residue 1 of the hydrolase; in Figure 14 Residues 9 and 16 of the isomerase are aligned with Residue 1 of the hydrolase. These observations are not readily reconciled with evolutionary divergence achieved by an overlay of random amino acid substitutions placed on a single ancestral sequence. Rather, it appears that oligonucleotides coding for segments of amino acid sequences were substituted from an ancestral isomerase gene into different sections of the diverging enol-lactone hydrolase gene.

The suggestion that oligonucleotide sequences may have been exchanged among genes coding for different enzymes of the β-ketoadipate pathway is fortified by observation of similar amino acid sequences in enzymes that subject chemically dissimilar substrates to different catalytic processes. Examples are protocatechuate oxygenase and β-ketoadipate enol-lactone hydrolase (Figure 3). The former enzyme introduces molecular oxygen across the double bond of a diphenol, and the latter enzyme cleaves a lactone ring with water. Despite the chemical dissimilarity of the reactions catalyzed by the enzymes, comparison of the amino acid sequences of the enzymes (Figure 15) reveals striking similarities; these cannot be attributed to the selection of similar catalytic or binding sites in the enzymes. Carboxymuconolactone decarboxylase mediates another biochemical type reaction, yet comparison of the NH_2-terminal amino acid sequence of this enzyme with the amino acid sequence of protocatechuate oxygenase reveals additional sequence similarities (Figure 16).

VIII. REPEATED AMINO ACID SEQUENCES OCCUR WITHIN THE PRIMARY STRUCTURE OF PROTEINS

Many proteins contain repeated amino acid sequences within their primary structures.[49] Often the sequence repetitions are attributed to mutations caused by single crossover events between misaligned homologous DNA sequences. As shown in Figure

Pseudomonas Oxygenase	69 GLU	70 VAL	71 TRP	72 GLN	-Δ-	73 ALA	74 ASP	75 ALA	76 ASP	77 GLY	78 GLU	79 TYR	80 GLN	81 ASP	82 ALA	83 TYR	84 ASN	85 LEU
Pseudomonas Hydrolase	1 ALA	2 HIS	3 LEU	4 GLN	5 LEU	6 ALA	7 ASP	8 GLY	9 VAL	10 LEU	11 ASN	12 TYR	13 GLN	-Δ-	14 ILE	15 ASP	16 GLY	17 PRO

Pseudomonas Oxygenase	86 GLU	87 ASN	88 ALA	-Δ-	-Δ-	-Δ-	-Δ-	-Δ-	89 PHE	90 ASN	91 SER	92 PHE	93 GLY	94 ARG	95 THR	96 ALA	97 THR	98 THR	99 PHE	100 ASP
Pseudomonas Hydrolase	18 GLU	19 ASN	20 ALA	21 PRO	22 VAL	23 LEU	24 VAL	25 LEU	26 SER	27 ASN	28 SER	29 LEU	30 GLY	31 THR	32 ASP	33 LEU	34 GLY	35 MET	36 ARG	37 ASP

FIGURE 15. Similar amino acid sequences in the primary structures of the α-subunit of *P. aeruginosa* protocatechuate 3,4-dioxygenase and *P. putida* β-ketoadipate enol-lactone hydrolase. Numerals above residues indicate their positions in the primary structures of the proteins. Identical residues are enclosed in boxes.

Oxygenase	23 LEU	24 GLU	25 ALA	26 ALA	27 GLY	28 ASN	29 PRO	30 THR	31 ARG	32 ASP	-Δ-	33 GLN	34 GLU	35 ILE	36 TRP	37 ASN	38 ARG	39 LEU	40 ALA	41 LYS	42 PRO	43 ASP
Decarboxylase	27 LEU	28 GLU	29 LYS	30 LEU	31 ASN	32 ASP	33 PHE	34 ASN	35 GLY	36 GLU	37 PHE	38 GLN	39 GLU	40 MET	41 ILE	42 THR	43 ARG	44 HIS	45 ALA	46 CYS	47 GLY	48 ASP

Oxygenase	54 GLN	55 VAL	56 TYR	57 ASP	58 GLY	59 ASP	60 GLY	61 HIS	62 LEU	63 VAL	64 ARG	65 ASP	66 SER	67 PHE	68 LEU	69 GLU	70 VAL	71 TRP	72 GLN	73 ALA	74 ASP	75 ALA	76 ASP	77 GLY	78 GLU
Decarboxylase	5 GLN	6 ARG	7 TYR	8 ASP	9 ALA	-Δ-	10 GLY	11 MET	12 GLN	13 VAL	14 ARG	15 ARG	16 ALA	17 VAL	18 LEU	19 GLY	20 ASP	21 ALA	22 HIS	23 VAL	24 ASP	25 ARG	26 SER	27 LEU	28 GLU

FIGURE 16. Similar amino acid sequences in the primary structures of the α-subunit of *P. aeruginosa* protocatechuate 3,4-dioxygenase and *P. putida* carboxymuconolactone decarboxylase. Numerals above residues indicate their positions in the primary structures of the proteins. Identical residues are enclosed in boxes.

17, the resulting mutant genes contain tandem duplications; comparison of the mutant and the parental gene requires introduction of a gap into the parental sequence (Figure 17). Sequence repetitions might also be caused by double crossover events between misaligned homologous sequences (Figure 18). Such mutations would give rise to sequence repetitions that need not be tandem; comparison of mutant and parental sequences would not require introduction of a gap.[48] Thus as shown in Figures 17 and 18, sequence repetitions caused by single crossover events may be distinguished from those caused by double crossover events by comparison of existing amino acid sequences with ancestral amino acid sequences; requirements for the introduction of gaps would indicate the occurrence of single crossover events. An opportunity to make this distinction is afforded by the NH_2-terminal amino acid sequences of muconolactone isomerases and *Pseudomonas* β-ketoadipate enol-lactone hydrolase. The sequences contain numerous internal repetitions (Figures 19 and 20), yet the repetitive patterns are different and therefore were introduced by genetic events unique to the evolutionary history of each protein. The *Pseudomonas* isomerase and *Pseudomonas* hydrolase share common ancestry with an amino acid sequence found in *Acinetobacter* enol-lactone hydrolase II (Figure 21). Few gaps are required to achieve the alignments depicted in Figures 19 to 21, and hence the internal sequence repetitions unique to the muconolactone isomerases and *Pseudomonas* enol-lactone hydrolase appear to have been superimposed by mutation upon a sequence represented in *Acinetobacter* enol-lactone hydrolase II. The data thus support the double crossover model in which repeated sequences are introduced by oligonucleotide substitution rather than by oligonucleotide insertion.

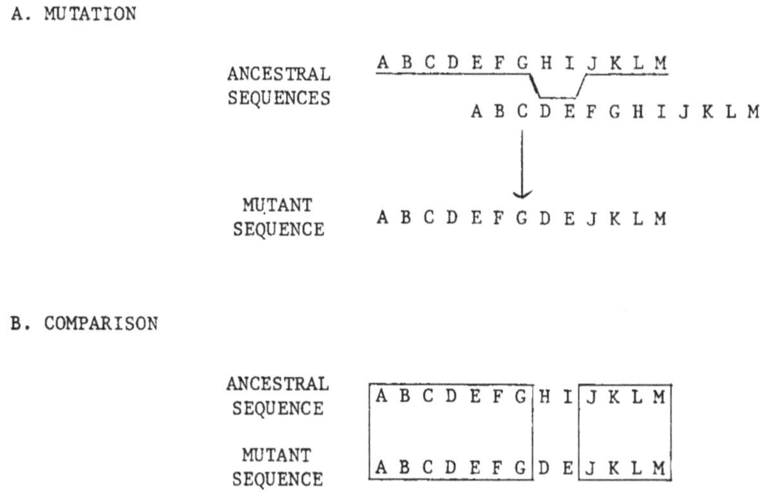

FIGURE 17. Repeated sequences caused by single crossover events between misaligned homologous DNA sequences. Recombination creates a mutant sequence containing tandem repetitions. Comparison of the mutant sequence and the ancestral sequence requires the introduction of a gap into the ancestral sequence.

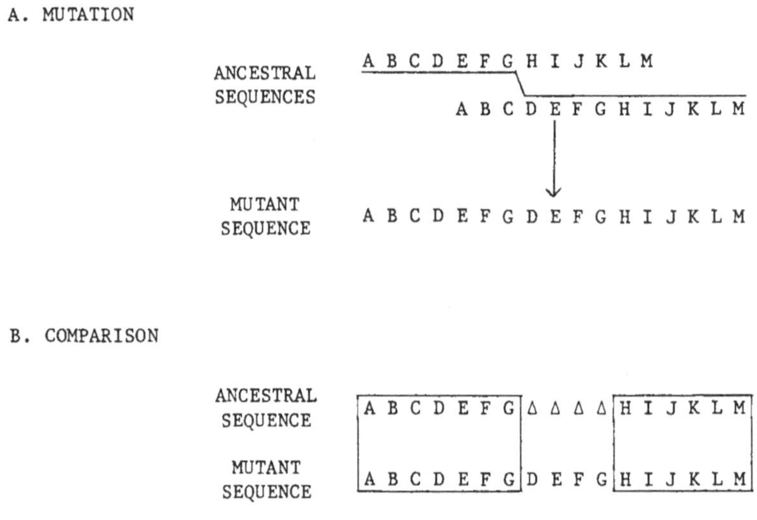

FIGURE 18. Repeated sequences caused by double crossover events between misaligned homologous DNA sequences. Repetitions created by the recombinational event need not be tandem. Comparison of the mutant sequence and the ancestral sequence does not require introduction of a gap into the ancestral sequence.

IX. INTERACTIONS AMONG CO-EVOLVING GENES: TOWARD MOLECULAR NATURAL HISTORY

We have reviewed data suggesting two kinds of relationships among co-evolving genes. One relationship, selective forces favoring the rapid divergence of lengthy DNA copies within a genome, is indicated by the dissimilar amino acid sequences of *Acinetobacter* enol-lactone hydrolases (Figure 9) and of carboxymuconolactone decarboxylases and muconolactone isomerases (Figure 12). The other relationship, an apparent

	1	2	3	4	5	6	7	8
Acinetobacter	MET	LEU	PHE	GLN	VAL	ARG	MET	ASP
Pseudomonas	MET	LEU	PHE	HIS	VAL	LYS	MET	THR

	9	10	11	12	13	14	15	16	17	18
Acinetobacter	VAL	HIS	LEU	PRO	VAL	SER	MET	PRO	THR	ASP
Pseudomonas	VAL	LYS	LEU	PRO	VAL	ASP	MET	ASP	PRO	ALA

	19	20	21	22	23	24	25	26	27
Acinetobacter	GLN	ALA	ASN	GLN	ILE	LYS	SER	VAL	GLU
Pseudomonas	LYS	ALA	THR	GLN	LEU	LYS	ALA	ASP	GLU

	28	29	30	31	32	33	34	35	36	37	38	39	40	41	42	43	44	45
Acinetobacter	LYS	ALA	TYR	SER	GLN	GLU	LEU	GLN	ARG	GLN	GLY	LYS	TRP	ARG	HIS	ILE	TRP	ARG
Pseudomonas	LYS	GLU	LEU	ALA	GLN	ARG	LEU	GLN	ARG	GLU	GLY	THR	TRP	ARG	HIS	LEU	TRP	ARG

	46	47	48	49	50	51	52	53	54	55
Acinetobacter	ILE	THR	GLY	GLN	TYR	ALA	ASN	ILE	SER	ILE
Pseudomonas	ILE	ALA	GLY	HIS	TYR	ALA	ASN	TYR	SER	VAL

	77	78	79	80	81	82	83	84	85	86	87	88
Pseudomonas	MET	ASX	ILE	ASX	GLY	LEU	CYS	ARG	HIS	PRO	SER	SER

	89	90		91	92	93	94	95	
Pseudomonas	ILE	ASX	-Δ-	SER	ASX	ARG	HIS	LYS	end

FIGURE 19. Internal sequence repetitions in the primary structures of *Pseudomonas* and *Acinetobacter* muconolactone isomerases. Numerals above residues indicate their positions in the primary structures of the proteins. Identical residues in the internally aligned sequences are enclosed in boxes. For the most part repeated sequence segments are not tandem. The amino acid sequence of the carboxy terminal fragment of *Pseudomonas* muconolactone isomerase was determined by Meagher.[50]

genetic mechanism for substitution of DNA oligonucleotide copies from one portion of the genome into another is illustrated by the appearance of similar amino acid sequences in enzymes with different catalytic activities (Figures 13 to 16). Intragenic sequence substitution is suggested by the internal repetitions found in the primary structure of muconolactone isomerases (Figure 19) and *Pseudomonas* β-ketoadipate enol-lactone hydrolase (Figure 20).

It is likely that additional evidence will illustrate the operation of other forces that direct the evolution of genes. For example, comparison of the structure of procaryotic and eucaryotic β-carbomuconate lactonizing enzymes (Figure 2) may reveal analogous protein structures shaped by convergent evolution; intrageneric comparison of regulatory regions that interact with inducers may indicate the extent to which genetic rearrangements were involved in the evolution of novel transcription controls. A full set of data may allow development of a molecular natural history of genes that is as rich and complex as better known natural histories of organisms.

ACKNOWLEDGMENTS

Research in our laboratory was supported by grants from the National Science Foundation (PCM 24884 and PCM 10075) and from the National Institutes of Health (GM 25487).

1	2	3	4	5	6	7	8	9	
ALA	HIS	LEU	GLN	LEU	ALA	ASP	GLY	VAL	
10	11	12	13	14		15	16	17	
LEU	ASN	TYR	GLN	ILE	-Δ-	ASP	GLY	PRO	
18	19	20	21	22	23	24	25	26	
GLU	ASN	ALA	PRO	VAL	LEU	VAL	LEU	SER	
27	28	29	30	31	32	33	34	35	
ASN	SER	LEU	GLY	THR	ASP	LEU	GLY	MET	
36	37	38	39	40	41		42	43	
ARG	ASP	THR	GLN	ILE	PRO	-Δ-	LEU	SER	
44	45	46	47		48	49	50	51	52
ASX	GLN	TYR	PHE	-Δ-	PRO	VAL	LEU	ARG	TYR
53	54								
ASX	GLN								

FIGURE 20. Internal sequence repetitions in the primary structure of *Pseudomonas* β-ketoadipate enol-lactone hydrolase. Numerals above residues indicate their positions in the primary structure of the protein. Identical residues in the internally aligned sequence are enclosed in boxes. For the most part repeated sequence segments are not tandem.

	1	2	3	4	5	6	7	8	9	10	11	12	13	14	15
Pseudomonas Isomerase	MET	LEU	PHE	HIS	VAL	LYS	MET	THR	VAL	LYS	LEU	PRO	VAL	ASP	MET
	1	2	3	4	5	6	7	8	9	10		11	12	13	14
Acinetobacter Hydrolase II	PRO	VAL	PHE	HIS	PHE	LYS	ASP	THR	LEU	THR	-Δ-	ALA	GLN	ASP	VAL
							1	2	3	4	5	6		7	8
Pseudomonas Hydrolase							ALA	HIS	LEU	GLN	LEU	ALA	-Δ-	ASP	GLY

	16	17	18	19	20	21	22	23	24	25	26	27	28	29	30
Pseudomonas Isomerase	ASP	PRO	ALA	LYS	ALA	THR	GLN	LEU	LYS	ALA	ASP	GLY	LYS	GLU	LEU
	15	16	17	18	19	20	21	22	23	24	25	26	27	28	29
Acinetobacter Hydrolase II	ALA	LEU	ASP	TYR	ALA	THR	PHE	GLY	GLN	ALA	ASP	ARG	PRO	ALA	LEU
	9	10	11	12	13	14	15	16	17	18	19	20	21	22	23
Pseudomonas Hydrolase	VAL	LEU	ASN	TYR	GLN	ILE	ASP	GLY	PRO	GLU	ASN	ALA	PRO	VAL	LEU

FIGURE 21. Similar amino acid sequences in the NH_2-terminal fragments of *Pseudomonas* muconolactone isomerase, *Acinetobacter* enol-lactone hydrolase II and *Pseudomonas* enol-lactone hydrolase. Numerals above residues indicate their positions in the primary structures of the proteins. Identical residues in the aligned sequences are enclosed in boxes.

REFERENCES

1. **Stanier, R. Y., Palleroni, N. J., and Doudoroff, M.,** The aerobic pseudomonads: a taxonomic study, *J. Gen. Microbiol.,* 43, 159, 1966.
2. **Ambler, R. P.,** The evolutionary stability of cytochrome-551 in *Pseudomonas aeruginosa* and *Pseudomonas fluorescens* biotype C, *Biochem. J.,* 137, 1, 1974.
3. **Stanier, R. Y. and Ornston, L. N.,** The β-ketoadipate pathway, in *Advances in Microbial Physiology,* Vol. 9, Rose, A. H. and Tempest, D. W., Eds., Academic Press, London, 1973, 89.

4. **Ornston, L. N. and Parke, D.**, Evolution of induction mechanisms in bacteria; insights derived from study of the β-ketoadipate pathway, in *Current Topics in Cellular Regulation*, Vol. 12, Horecker, B. L. and Stadtman, E. R., Eds., Academic Press, New York, 1977, 209.
5. **Cain, R. B., Bilton, R. F., and Darrah, J. A.**, The metabolism of aromatic acids by microorganisms. Metabolic pathways in the fungi, *Biochem. J.*, 108, 797, 1968.
6. **Rann, D. L. and Cain, R. B.**, Regulation of the enzymes of aromatic ring fission in the genus *Nocardia, Biochem. Soc. Trans.*, 1, 658, 1969.
7. **Ornston, L. N. and Stanier, R. Y.**, The conversion of catechol and protocatechuate to β-ketoadipate by *Pseudomonas putida*. I. Biochemistry, *J. Biol. Chem.*, 241, 3776, 1966.
8. **Thatcher, D. R. and Cain, R. B.**, The metabolism of aromatic acids by microorganisms: the conversion of 3-carboxymuconolactone to 3-oxoadipate in *Aspergillus niger, Biochem. J.*, 120, 28, 1970.
9. **Ornston, L. N.**, The conversion of catechol and protocatechuate to β-ketoadipate by *Pseudomonas putida*. II. Enzymes of the protocatechuate pathway, *J. Biol. Chem.*, 241, 3787, 1966.
10. **Thatcher, D. R. and Cain, R. B.**, Molecular weight and physical properties of 3-carboxy-*cis,cis*-muconate cyclase, *Eur. J. Biochem.*, 48, 549, 1974.
11. **Thatcher, D. R. and Cain, R. B.**, The subunit structure of 3-carboxy-*cis,cis*-muconate cyclase, *Eur. J. Biochem.*, 48, 557, 1974.
12. **Patel, R. N., Meagher, R. B., and Ornston, L. N.**, Relationships among enzymes of the β-ketoadipate pathway. II. Properties of crystalline β-carboxy-*cis,cis*-muconate lactonizing enzyme from *Pseudomonas putida, Biochemistry*, 12, 3531, 1973.
13. **Kohlmiller, N. A. and Howard, J. B.**, The primary structure of the α-subunit of protocatechuate 3,4-dioxygenase. II. Isolation and sequence of overlap peptides and complete sequence, *J. Biol. Chem.*, 254, 7309, 1979.
14. **Iwaki, M., Kagamiyama, H., and Nozaki, M.**, The complete amino acid sequence of the β-subunit of protocatechuate 3,4-dioxygenase from *Pseudomonas aeruginosa, J. Biochem., (Tokyo)*, 86, 1159, 1979.
15. **Hardisson, C., Sala-Trepat, J., and Stanier, R. Y.**, Pathways for the oxidation of aromatic compounds in *Azotobacter, J. Gen. Microbiol.*, 59, 1, 1969.
16. **Sala-Trepat, J. M. and Evans, W. C.**, The meta cleavage of catechol by *Azotobacter* species. 4-Oxalocrotonate pathway, *Eur. J. Biochem.*, 20, 400, 1971.
17. **Dagley, S.**, Pathways for the utilization of organic growth substrates, in *The Bacteria*, Vol. 6, Ornston, L. N. and Sokatch, J. R., Eds., Academic Press, New York, 1978, 305.
18. **Robert-Gero, M., Poiret, M., and Stanier, R. Y.**, The function of the β-ketoadipate pathway in *Pseudomonas acidovorans, J. Gen. Microbiol.*, 57, 207, 1969.
19. **Ornston, M. K. and Ornston, L. N.**, The regulation of the β-ketoadipate pathway in *Pseudomonas acidovorans* and *Pseudomonas testosteroni, J. Gen. Microbiol.*, 73, 455, 1972.
20. **Ornston, L. N.**, The conversion of catechol and protocatechuate to β-ketoadipate by *Pseudomonas putida*. IV. Regulation, *J. Biol. Chem.*, 241, 3800, 1966.
21. **Meagher, R. B., McCorkle, G. M., Ornston, M. K., and Ornston, L. N.**, Inducible uptake system for β-carboxy-*cis,cis*-muconate in a permeability mutant of *Pseudomonas putida, J. Bacteriol.*, 111, 465, 1972.
22. **Ingledew, W. M., Tresguerres, M. E. F., and Canovas, J. L.**, Regulation of the enzymes of the hydroaromatic pathway in *Acinetobacter calcoaceticus, J. Gen. Microbiol.*, 68, 273, 1971.
23. **Canovas, J. L. and Stanier, R. Y.**, Regulation of the enzymes of the β-ketoadipate pathway in *Moraxella calco-acetica*. I. General aspects, *Eur. J. Biochem.*, 1, 289, 1967.
24. **Johnson, B. F. and Stanier, R. Y.**, Regulation of the β-ketoadipate pathway in *Alcaligenes eutrophus, J. Bacteriol.*, 107, 476, 1971.
25. **Durham, D. R. and Ornston, L. N.**, Homologous structured genes and similar induction patterns in *Azotobacter* and *Pseudomonas, J. Bacteriol.*, 143, 834, 1980
26. **Crawford, I. P.**, Gene rearrangements in the evolution of the tryptophan synthetic pathway, *Bacteriol. Rev.*, 39, 87, 1975.
27. **Fujisawa, H. and Hayaishi, O.**, Protocatechuate 3,4-dioxygenase. I. Crystallization and characterization, *J. Biol. Chem.*, 243, 2673, 1968.
28. **Yoshida, R., Hori, K., Fujiwara, M., Saeki, Y., Kagamiyama, H., and Nozaki, M.**, Nonidentical subunits of protocatechuate 3,4-dioxygenase, *Biochemistry*, 15, 4048, 1976.
29. **Wells, M.**, Ph.D. dissertation, University of Texas, Austin, 1972.
30. **Bull, C., Ballou, D. P., and Salmeen, I.**, Raman spectrum of protocatechuate oxygenase from *Pseudomonas putida*. New low frequency bands, *Biochem. Biophys. Res. Commun.*, 87, 836, 1979.
31. **Durham, D. R., Stirling, L. A., Ornston, L. N., and Perry, J. J.**, Intergeneric evolutionary homology revealed by the study of protocatechuate 3,4-dioxygenase from *Azotobacter vinelandii, Biochemistry*, 19, 149, 1980.

32. **Que, L., Lipscomb, J. D., Munck, E., and Wood, J. M.**, Protocatechuate 3,4-dioxygenase, Inhibitor studies and mechanistic implications, *Biochim. Biophys. Acta,* 485, 60, 1977.
33. **Parke, D.**, Structural comparison of γ-carboxymuconolactone decarboxylase and muconolactone isomerase from *Pseudomonas putida, Biochim. Biophys. Acta,* 578, 145, 1979.
34. **Yeh, W. K., Fletcher, P., and Ornston, L. N.**, Homologies in the NH_2-terminal amino acid sequences of γ-carboxymuconolactone decarboxylases and muconolactone isomerases, *J. Biol. Chem.,* 255, 6347, 1980.
35. **Yeh, W. K., Durham, D. R., Fletcher, P., and Ornston, L. N.**, Evolutionary relationships among carboxymuconolactone decarboxylases, *J. Bacteriol.,* 146, 233, 1981.
36. **Patel, R. N., Meagher, R. B., and Ornston, L. N.**, Relationships among enzymes of the β-ketoadipate pathway. IV. Muconolactone isomerase from *Acinetobacter calcoaceticus* and *Pseudomonas putida, J. Biol. Chem.,* 249, 7410, 1974.
37. **Yeh, W. K., Davis, G., Fletcher, P., and Ornston, L. N.**, Homologous amino acid sequences in enzymes mediating sequential metabolic reactions, *J. Biol. Chem.,* 253, 4920, 1978.
38. **Stanier, R. Y., Wachter, D., Gasser, C., and Wilson, A. C.**, Comparative immunological studies of two *Pseudomonas* enzymes, *J. Bacteriol.,* 102, 351, 1970.
39. **Hou, C.-T. and Lillard, M. O.**, Immunological properties of protocatechuate 3,4-dioxygenase isofunctional enzymes, *J. Bacteriol.,* 126, 516, 1976.
40. **Patel, R. N., Hou, C.-T., Felix, A., and Lillard, M. O.**, Catechol 1,2-dioxygenase from *Acinetobacter calcoaceticus*: Purification and properties, *J. Bacteriol.,* 127, 536, 1976.
41. **Patel, R. N. and Ornston, L. N.**, Immunological comparison of enzymes of the β-ketoadipate pathway, *Arch. Microbiol.,* 110, 27, 1976.
42. **Patel, R. N., Mazumdar, S. and Ornston, L. N.**, β-Ketoadipate enol-lactone hydrolases I and II from *Acinetobacter calcoaceticus, J. Biol. Chem.,* 250, 6567, 1975.
43. **Yeh, W. K., Fletcher, P., and Ornston, L. N.**, Evolutionary divergence of coselected β-ketoadipate enol-lactone hydrolases in *Acinetobacter calcoaceticus, J. Biol. Chem.,* 255, 6342, 1980.
44. **McCorkle, G. M., Yeh, W. K., Fletcher, P., and Ornston, L. N.**, Repetitions in the NH_2-terminal amino acid sequence of β-ketoadipate enol-lactone hydrolase from *Pseudomonas putida, J. Biol. Chem.,* 255, 6335, 1980.
45. **Ornston, L. N. and Parke, D.**, Evolution of catabolic pathways, *Biochem. Soc. Trans.,* 4, 468, 1976.
46. **Haber, J. E. and Koshland, D. E.**, An evaluation of the relatedness of proteins based on comparison of amino acid sequence, *J. Mol. Biol.,* 50, 617, 1970.
47. **Meagher, R. B. and Ornston, L. N.**, Relationships among enzymes of the β-ketoadipate pathway. I. Properties of *cis,cis*-muconate lactonizing enzyme and muconolactone isomerase from *Pseudomonas putida, Biochemistry,* 12, 3523, 1973.
48. **Ornston, L. N. and Yeh, W. K.**, Origins of metabolic diversity: evolutionary divergence by sequence repetition, *Proc. Natl. Acad. Sci., U.S.A.,* 76, 3996, 1979.
49. **Barker, W. C., Ketcham, L. K., and Dayhoff, M. O.**, A comprehensive examination of protein sequences for evidence of internal gene duplication, *J. Mol. Evol.,* 10, 265, 1978.
50. **Meagher, R. B.**, Purification and partial amino acid sequence of the cyanogen bromide fragments of muconolactone isomerase from *Pseudomonas putida, Biochim. Biophys. Acta,* 494, 33, 1977.

Chapter 6

GENETIC MECHANISMS IN THE DISSIMILATION OF CHLORINATED COMPOUNDS

A. M. Chakrabarty

TABLE OF CONTENTS

I. INTRODUCTION

We live in an era where chemicals play a vital part in our everyday life. Large quantities of various chemicals, some of which are highly toxic, are used routinely by industries and consumers either for material comfort or for enhanced agricultural productivity and manufacture of industrial products. This has led to a tremendous growth of the chemical industry in the U.S. as well as in other countries of the world. Production levels of synthetic organic chemicals have doubled every 8 years for the last 3 decades in the U.S. alone.

One of the problems of the extensive use of chemicals is the release of large number of various toxic chemicals in the environment. The debate about the health risks associated with industrial chemicals has continued for some time[1] and the public concern over the increasing incidences of cancer due to environmental carcinogens has increased substantially and led to tighter governmental regulations of the production and use of many forms of toxic chemicals.[2] Incidences such as the dumping of polychlorinated biphenyls in the Hudson River in New York or Kepone® in the James River in Virginia or the accidental contamination of livestock feeds with polybrominated biphenyls (PBBS) in Michigan, have all added to public anxiety over the large-scale usage of highly halogenated compounds. The contamination of soil and ground water due to slow seepage of buried chemicals from waste dump sites, as evidenced by the Love Canal incident in New York, again has drawn attention to the long-term public health concern over the methods employed for storing radioactive and toxic chemicals. A glimpse of the magnitude of the problem can be obtained from the fact that 762 million tons of chemical waste generated by 53 largest U.S. chemical companies are located in 3383 dumping sites all over the country.[3]

One of the major problems in the use of the chemicals, aside from toxicity and carcinogenicity of a large proportion of them,[4] is their persistence in nature. The extremely long life of such chemicals, when present in the soil or the aquatic environment, greatly amplifies the toxicity and health risk problems in the area of contamination. This, of course, is a reflection of the inability of the local microflora to dissimilate such compounds. Traditionally, microorganisms have been responsible for maintaining the carbon, nitrogen, and sulfur balances in the environment, and the inability of the soil or aquatic microorganisms to mineralize the halogenated and other synthetic chemicals, therefore, adds to the toxic load of the soil.

Microorganisms are, of course, well known for their adaptability to extreme environments; although the initial response has been slow, there is now increasing evidence that microorganisms are "learning" to cope with the multitude of toxic, man-made chemicals that have been released into the environment in massive amounts.[5] The dissimilation of polychlorinated biphenyls by soil bacteria and the mechanism of anaerobic and aerobic dechlorination from the aromatic and aliphatic compounds are the subject matters presented by other investigators elsewhere in this book. In this article, I would like to review some aspects of microbial dissimilation of chlorinated compounds, with particular emphasis on the role of plasmids in such processes, and genetic mechanisms governing dissimilation of chlorinated compounds by members of Enterobacteriaceae.

II. MICROBIAL DISSIMILATION OF CHLORINATED COMPOUNDS

Whether or not a chlorinated compound will be utilized by microorganisms may depend upon two factors: the structural similarity of the compound with a natural product or the ability of such compounds to induce the synthesis of enzymes that may act on structurally analogous, naturally occurring compounds. Thus DDT is seldom acted on by microorganisms, but if the *para* chlorine atoms of DDT are replaced by methoxyl groups, resembling natural products of similar structure, the resulting compound (methoxychlor) is much more amenable to microbial attack.[6] Thus the number and the position of chlorine atoms in the molecule determine, to a large extent, the biodegradability of the compound.[7]

It should be emphasized that although limited in number, there are natural compounds that contain carbon-halogen bonds[8] so that microorganisms may have been exposed to at least some form of halogenated compounds before the introduction of man-made compounds in the environment. To what extent such an exposure may have affected the evolution of dehalogenation functions against the man-made compound is unknown at present. There is, however, ample evidence that many highly chlorinated compounds are reported to be biodegraded even by pure cultures. Furukawa, chapter 2, has reviewed the extent of microbial degradation of polychlorinated biphenyls by pure and mixed cultures, while the biodegradability of various herbicides and pesticides has been reviewed.[10] Pure cultures capable of utilizing various chlorinated compounds, including chlorinated aromatics, have been isolated and in many cases, the pathway of degradation has been studied.[11,12] More interestingly, pure cultures have been described that can utilize as a sole source of carbon, a highly chlorinated aromatic compound such as pentachlorophenol.[13-15] Single cultures such as *Pseudomonas* sp. or *Corynebacterium* strain KC-3 have been reported to utilize pentachlorophenol, 2,3,4,6-tetrachlorophenol, and 2,4,6-trichlorophenol as growth substrates.[14,15] These and other chlorophenols are metabolized to various degrees by the cultures. The mechanisms of breakdown of pentachlorophenol has been shown to involve its conversion to partially dechlorinated hydroquinone intermediates, which then undergo ring cleavage. The critical intermediates have been characterized to be 2,6-dichloroquinone and 2,3,5,6-tetrachlorohydroquinone or 2,3,5,6-tetrachlorobenzohydroquinone.[15] The ability of pure cultures to cleave chlorine atoms from the aromatic nucleus prior to ring cleavage has therefore been documented. The ability of mixed cultures to dechlorinate 1,2,3- and 1,2,4-trichlorobenzene with the formation of 2,3-dichlorophenol and 2,4-dichlorophenol and subsequent hydroxylation and ring cleavage has also been demonstrated.[16]

III. SPECIFICITY AND EVOLUTION OF NEW ENZYMES IN THE BIODEGRADATION OF CHLORINATED COMPOUNDS

The presence of chlorine atoms on the aromatic nucleus is known to greatly retard the chemical reactivity, particularly towards the electrophilic form of oxygen which is involved before ring cleavage can take place. This reduced reactivity must have been responsible for a slow rate of degradation of most chlorinated aromatic compounds. Many enzymes capable of replacing a chlorine atom by OH^- from water in case of chlorinated aliphatic compounds are widely known.[17,18] Thus while simple chlorinated aliphatic acids have been known to be biodegradable, the chlorinated aromatic compounds in general appear to be only slowly degraded, presumably by co-oxidative processes.[19,20]

It now appears that microorganisms are evolving new enzymes that can act on the chlorinated compound, but not, or only slowly, on the nonchlorinated parent. Thus Crawford et al.[12] have isolated a strain of *Bacillus brevis* that can utilize 5-chlorosalicylate as a sole source of carbon and energy. This bacterium synthesizes a specific 5-chlorosalicylate 1,2-dioxygenase, a novel enzyme that cleaves the aromatic ring while it is substituted with only one hydroxyl group (normally two hydroxyl groups, oriented *ortho* or *para* to each other are required in the aromatic nucleus before ring cleavage). Interestingly, this novel enzyme cannot use salicylate as a substrate. A halogen other than iodine or a methoxyl group is required at the 5 position for a substitute salicylate to be a substrate for the dioxygenase. The loss of the halogen substituent as halide ion from ring carbon 5 also appears to be enzymatically mediated, although the substrate specificity of this enzyme could not be studied because of the unavailability of suitable substrates.

Knackmuss and associates[21,22] have studied the biochemistry of 3-chlorobenzoate degradation by a *Pseudomonas* sp. B3. They demonstrated the stereospecific nature of the initial benzoate 1,2-dioxygenase enzyme that could act on 3-chlorobenzoate but not on 4-chloro or 3,5-dichlorobenzoates. Two separate catechol 1,2-dioxygenases have been characterized in the 3-chlorobenzoate-grown cells of strain B3.[23] Introduction of the TOL plasmid, which codes for a benzoate 1,2-dioxygenase of a broad substrate specificity allows selection of B3 cells that could also utilize 4-chloro- and 3,5-dichlorobenzoates.[11,24] Thus the chlorocatechol metabolism is accomplished by isoenzymes that appear to have a broader specificity. The nature of enzymes allowing release of chloride ions from the chlorobenzoate substrates has not been studied in detail. It is likely that some of the isoenzymes involved in the 3-chlorobenzoate degradative pathway have specifically evolved for the oxidative degradation of this compound by *Pseudomonas* sp. B3.

Genetic evidence for the evolution of new enzymes responsible for oxidative dissimilation of 3-chlorobenzoate has been obtained by transferring a transmissible plasmid (pAC25) coding for the complete degradation of 3-chlorobenzoate. When pAC25 was transferred to a number of nonallelic benzoate-negative (Ben^-) or *p*-hydroxybenzoate-negative (Pob^-) mutants of *P. aeruginosa* or *P. putida*, all the exconjugants could grow well with 3-chlorobenzoate, but did not regain the ability to utilize benzoate or hydroxybenzoates.[25] No functional complementation was observed in presence of inducing quantities of 3-chlorobenzoate, suggesting that the plasmid codes for some enzymes that are specifically involved in the degradation of 3-chlorobenzoate and some of its metabolites.

IV. PLASMIDS AND BIODEGRADATION OF HYDROCARBONS, CHLORINATED HYDROCARBONS, AND THEIR OXIDATIVE METABOLITES

The evolution of completely new sets of enzymes that can specifically act on chlorinated compounds as substrates, but not, or only slowly, the nonchlorinated parents, is immensely important, since it raises the interesting question as to how microorganisms evolve new enzymatic functions and how such a process of evolution can be accelerated for enhanced biodegradation of various synthetic environmental pollutants. It is now well known that the genes specifying biodegradation of a number of chlorinated and nonchlorinated hydrocarbons are borne on plasmids, rather than the chromosome of the microorganisms.[25,26] A list of such plasmids is given in Table 1.

The occurrence of degradative pathways for toxic hydrophobic and chlorinated compounds such as octane, xylene, toluene, *p*-chlorobiphenyl, etc. in the form of plasmids raises the question whether such plasmids evolve, in analogy with antibiotic resistance, as a primary means of detoxification.[26] The presence of chlorine atoms in organic molecules not only contributes to their toxicity towards animals, but also to microorganisms. The transmissible plasmid nature of the genes, specifying dissimilation of such compounds, may therefore lead to a rapid spread of the degradative capabilities against such toxic man-made compounds under conditions of high localized concentration of the compounds. If so, one would expect that the mechanism of evolution of degradative plasmids would be similar to that of antibiotic-resistance plasmids. A

Table 1
LIST OF DEGRADATIVE PLASMIDS

Plasmid	Degradative pathway	Transmissibility[a]	Size (× 10⁶)
CAM	Camphor	+	>100[b]
OCT	*n*-Octane	−	>100
SAL	Salicylate	+	55,48,42
NAH	Naphthalene	+	46
TOL	Xylene/toluene	+	76
XYL-K	Xylene/toluene	+	90
2-HP	2-Hydroxypyridine	?	63
NIC	Nicotine/nicotinate	+	ND[c]
pOAD2	6-Aminohexanoic acid cyclic dimer	−	29
pJP1	2,4-Dichlorophenoxy-acetic acid	+	58
pAC8 (RP4-TOL)	Xylene/toluene	+	76
pAC21	*p*-Chlorobiphenyl	+	65
pAC25	3-Chlorobenzoate	+	68

[a] Transmissible plasmids are denoted as +, nontransmissible ones as −.
[b] Exact size unknown.
[c] ND, not determined.

glimpse of the similarity in the mechanism of evolution of these two types of plasmids comes from the fact that some degradative and antibiotic-resistance plasmids, belonging to incompatibility group P9, show significant amount of genetic homology, as measured by restriction hybridization.[27] Thus the degradative plasmid TOL exhibits about 18 to 20% homology with the antibiotic-resistance plasmid pMG18 and 12 to 14% with R2.

Apart from genetic homology, degradative plasmids demonstrate some properties that are characteristics of antibiotic-resistance plasmids. Thus transmissible degradative plasmids often dissociate into nontransmissible and sex factor plasmids when transferred to a different host,[28,29] similar to the dissociation of some antibiotic-resistance plasmids on being transferred from *Escherichia coli* to *Proteus mirabilis*.[30] Evidence is also accumulating that suggests that many antibiotic-resistance plasmids contain gene segments that code for resistance to antibiotics and that can be transposed from one replicon to another via *recA*-independent mechanisms.[31]

Translocation of segments of hydrocarbon degradative plasmids such as TOL onto sex factor or antibiotic-resistance plasmids has been reported.[29,32,33] There appear to be two specific segments of the TOL plasmid that are involved in specific excision and translocation. One is a 28-Mdalton segment containing all or part of the toluene degradative genes that is specifically excised from the TOL plasmid on growth of the TOL$^+$ host cells with benzoate.[34] A similar 28-Mdalton segment containing all the toluene degradative genes dissociates from the TOL plasmid to form a nontransmissible TOL* plasmid, when TOL is transferred from *P. putida* to *P. aeruginosa*.[29] A second segment is a 38-Mdalton fragment of TOL that can readily recombine with antibiotic-resistance plasmids such as RP4.[29,32,33] The recombinant RP4-TOL plasmid (pAC8) can afford resistance to Cb and Km, but not to Tc, suggesting that the recombinational site is at or near the Tc resistance gene leading to the inactivation of this gene.[29,33] That the 38-Mdalton TOL fragment also contains the 28-Mdalton TOL* fragment[27] is seen from

Table 2
BENZOATE-INDUCED EXCISION OF TOL$^+$ CHARACTER IN *recA*$^+$ AND LACK OF EXCISION IN *recA* HOSTS

Strain	Plasmid	Size (Mdalton)	Number of growth cycles in benzoate	Percent loss	Size of residual plasmid (Mdalton)
AC137	TOL	76	2	91	48
AC797	K-TOL*	110	2	71	84
AC810	pAC8	76	4	81	48
AC814[a]	TOL	76	2	<1	76
AC815[a]	pAC8	76	5	<1	76

[a] AC814 and AC815 are *P. putida* harboring *trp B615* and *recA*801 mutations.[35] Spontaneous loss of most of these plasmids is about 0.1 to 1% under similar growth conditions (glucose-grown).

the fact that growth of cells harboring pAC8 with benzoate leads to production of TOL$^-$ derivatives, which harbor residual plasmids of an average size of 48 million daltons (Table 2). It has also been possible to recombine TOL* segment with sex factors such as K, and excision of TOL* when host cells harboring K-TOL* are grown with benzoate leads to production of K$^+$ cells that have lost the entire TOL* segment (Table 2).

The excision of the 28-Mdalton TOL* fragment from pAC8 (RP4-TOL) under conditions of benzoate growth leaving behind a 10-Mdalton TOL fragment on RP4 raises the interesting question whether specific nucleotide sequences are involved in the excision of the 28-Mdalton fragment. Insertion sequences such as IS1 or IS3 have been implicated in the *recA*-independent association-dissociation or translocation of antibiotic-resistance plasmid segments in some cases.[30,31] It was, therefore, of interest to find out if excision of the TOL* fragment is dependent in the host recombination functions. When plasmids such as TOL or pAC8 were introduced into *P. putida* cells harboring *recA801* mutations[35] and such cells were grown with benzoate, there was no excision of the TOL* segment (Table 2). The excision process is therefore dependent on host *recA*$^+$ functions, similar to that of F from the F′ ORF203, which occurs presumably by recombination at the directly repeated insertion sequence 2 (IS2) in *E. coli.*[36] Similar to F′ ORF203, *recA*-dependent recombination at direct repeats may be involved in the excision of the TOL* segment.[27]

V. EXPRESSION OF HYDROCARBON DEGRADATIVE GENES IN ENTERIC AND OTHER BACTERIA

Screening of bacterial cultures from soil or aquatic samples contaminated with waste or crude oil for their ability to degrade hydrocarbons usually demonstrates this ability to be confined to certain aerobic bacterial genera, fungi, and some actinomyces. The vast majority of bacterial genera, particularly those belonging to the family Enterobacteriaceae, are not known for their ability to utilize hydrocarbons. Enteric bacteria such as *Klebsiella* or *Serratia* have soil and water as their natural habitats, similar to that of *Pseudomonas.* Since hydrocarbon degradative genes occur in the form of transmissible plasmids in *Pseudomonas* (Table 1), it is surprising that such abilities have not been passed along to their enteric neighbors. It was, therefore, of interest to see if enteric and other bacteria, normally incapable of hydrocarbon utilization, have any inherent defect in expressing hydrocarbon degradative genes. The transposition of hydrocarbon degradative gene segments such as TOL* (specifying xylene/toluene utilization) or SAL* (salicylate utilization) onto broad-host-range antibiotic-resistance plasmids such as PR4 or R702 has enabled us to introduce the hydrocarbon degradative gene segments into a number of enteric, soil, and aquatic bacteria such as *E. coli, Salmonella typhimurium, Agrobacterium tumefaciens, Azotobacter vinelandii, Serratia marcescens,* etc. Interestingly, none of these bacteria, which are normally incapable of utilizing hydrocarbons, can fully express the hydrocarbon degradative gene segments and grow with the appropriate hydrocarbons as the sole source of carbon (i.e., with xylene/toluene or salicylate). Their inability to express the degradative genes is not due to loss of the gene segments, since such RP4::TOL* or RP4::SAL* plasmids can be efficiently transferred from *E. coli, Agrobacterium* or *Serratia* species to *Pseudomonas* sp., where the TOL* or SAL* gene segments would be fully functional.[29] Extensive assays for the TOL*-specified enzymes in the laboratory of D.W. Ribbons at the University of Miami have indicated the presence of all the *meta* pathway enzymes needed for conversion of toluene to pyruvate and acetate, in RP4::TOL*[+] *E. coli,* although at a low level (10 to 25% of *P. putida* levels; see Reference 37). The presence of low levels of all the TOL enzymes in *E. coli* suggests that the lack of growth of *E. coli* with toluene is not due to the absence of any critical enzyme of the *meta* pathway, but rather due to the absence of some unknown factors needed for full expression of such genes in these microorganisms. It is presently assumed that such factors are cell envelope proteins present in pseudomonads and probably in other bacterial species capable of oxidation of hydrophobic compounds and are coded by the chromosomal genes. Lack of expression may also be due to a lack of recognition of the plasmid promoter sequences by the enteric RNA polymerases.

VI. EVOLUTION OF NEW METABOLIC COMPETENCE FOR HYDROCARBON OXIDATION IN THE FAMILY ENTEROBACTERIACEAE

It seems that the lack of the ability of members of the family Enterobacteriaceae (*E. coli, Salmonella, Klebsiella, Serratia,* etc.) to utilize hydrocarbons is not so much because of the absence of appropriate hydrocarbon degradative genes in these bacteria, but because of an inherent inability of the bacteria to express such genes when they are present. Why certain bacterial genera such as *Pseudomonas* or *Acinetobacter* evolved efficient hydrocarbon degradative competence, while others did not, remains a mystery. It is, however, clear that a lack of expression of such genes in various soil and aquatic bacteria must have contributed toward the slow pace of evolution of degradative functions against man-made toxic pollutants as compared to evolution of antibiotic-resistance functions which are basically expressed in most microorganisms including those transferred from Gram-positive to Gram-negative bacteria,[38] although occasional lack of expression of certain antibiotic-resistance genes is known. It should be stressed that gene segments, encoding specialized functions in the natural ecosystem, are known not to be expressed when transferred to microorganisms normally incapable of performing such specialized functions. Thus Ti-plasmid DNA segments, specifying catabolism of octopine or induction of crown-gall tumors on plants, are not expressed in *E. coli* when transferred from *A. tumefaciens,* which can fully express such genes.[39]

There are two very strong selective pressures that are acting on soil and aquatic microflora including members of Enterobacteriaceae to evolve new aromatic hydrocarbon degradative functions. One is the release of large quantities of toxic chlorinated hydrocarbons. The other is the fact that such compounds, particularly Kepone® and PCB type of compounds, are accumulating in the river sediments where the environment is basically anaerobic. There is an estimated 450 million pounds of PCBs in the soil and river sediments,[40] along with millions of pounds of other chlorinated hydrocarbons. Since anaerobes and facultative anaerobes comprise the vast majority of river sediment microbiota, there must have been strong selective pressure on them to detoxify such compounds. Yet, biodegradation of hydrocarbons is known to need participation of molecular oxygen, so that facultative anaerobes will have a better chance of biodegradation of hydrocarbons under conditions of appropriate oxygen tension. We have, therefore, been interested in isolating facultative anaerobes such as *Klebsiella* and *Serratia* species from the PCB-contaminated Hudson River sediments and examining the evolution of PCB and other hydrocarbon degradative competence in such strains. Indeed, as shown in Table 1, we have characterized a PCB plasmid in the *Klebsiella* species that allows slow growth with *p*-chlorobiphenyl.[41] Interestingly, this *Klebsiella* strain grows very well with other aromatic hydrocarbons such as benzoate and *p*-hydroxybenzoate. We have also isolated a *Serratia* species that grows slowly with *p*-chlorobiphenyl and also with benzoate, but not with *p*-hydroxybenzoate.

The ability of the *Klebsiella* or the *Serratia* species to grow with PCB or benzoate suggests that new metabolic functions involving biodegradation of PCB and other aromatic hydrocarbons are evolving in facultative anaerobes present in the PCB-contaminated sediments of the Hudson River. We have also recently demonstrated[41] that introduction of the RP4-TOL plasmid (pAC8) to such *Klebsiella* or *Serratia* species elicits selection of TOL^+ cells that can express functionally the TOL genes and grow at the expense of xylene or toluene. Since laboratory strains of enteric bacteria are totally incapable of hydrocarbon utilization, while the presence of the PCB-degradative plasmid (pAC21) allows some mutant cells to utilize hydrocarbons, it is clear that the lack of hydrocarbon oxidation by enteric bacteria is due to absence of more than one enzyme. The nature of enzymatic functions, coded by pAC21 to allow host cells to functionally express TOL genes, is unknown at present.

In order to determine if the gene(s), allowing expression of toluene degradative genes, evolved as part of pAC21 plasmid, or separately on the chromosome, we have transferred pAC8 to both $pAC21^+$ and $pAC21^-$ *K. pneumoniae* cells, and examined the ability of such cells to functionally express TOL genes and grow with toluene as a sole source of carbon and energy. The results are shown in Figure 1.

As can be seen, the presence of pAC21 is essential for allowing the host *K. pneumoniae* cells to grow on plates containing toluene as a sole source of carbon; pAC21-negative segregants are incapable of utilizing the hydrocarbon. Thus the expression genes must have evolved as part of pAC21 plasmid.

VII. EVOLUTION OF DECHLORINATION FUNCTIONS IN THE FORM OF PLASMIDS

There are three lines of evidence that suggest that the *K. pneumoniae* strain AC901 can convert *p*-chlorobiphenyl to *p*-chlorobenzoic acid, but not beyond this state. As mentioned previously,[41] growth of *K. pneumoniae* AC901 with *p*-chlorobiphenyl leads to an accumulation of *p*-chlorobenzoic acid, which can be detected by gas-chromatography — mass spectrometry. AC901 is also incapable of utilizing *p*-chlorobenzoic acid as sole source of carbon and energy. Thirdly, there is no liberation of chloride ions during growth of AC901 with *p*-chlorobiphenyl. All the evidence, therefore, points to the fact that the PCB degradative plasmid (pAC21) specifies an incomplete pathway of the bioconversion of *p*-chlorobiphenyl to *p*-chlorobenzoic acid.

Since the 2,4-D degradative plasmid (pJP1) codes for the partial conversion of 2,4-dichlorophenoxyacetic acid to 2,4-dichlorophenol,[42] it might be argued that plasmids encoding dechlorination steps have not yet evolved in nature.

We have recently characterized a 68-Mdalton transmissible plasmid pAC25 that codes for complete biodegradation of 3-chlorobenzoate (3 Cba, see References 25 and 43). As can be seen from the results in Figure 2 that during growth of the host cells

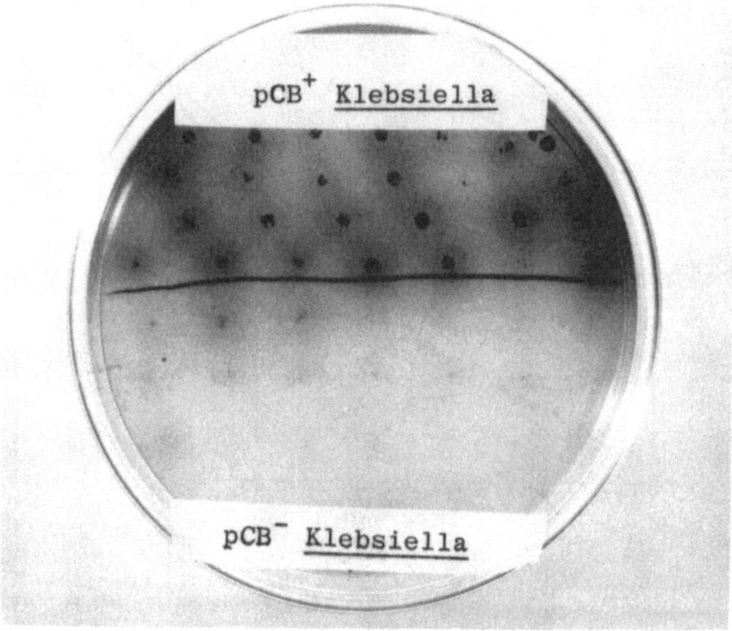

FIGURE 1. The ability of pCB+ *Klebsiella* (pAC21-positive) to grow with toluene as a sole source of carbon. The lower part of the plate demonstrates the inability of pCB- *Klebsiella* cells (pAC21-negative) to utilize toluene. Both strains harbor pAC8 plasmid.

harboring pAC25, almost 80% of the chlorine is released in the form of chloride ions in 48 hr. Loss of the plasmid from the cells leads to a loss of the ability of the cells to utilize 3-Cba and release chloride ions. In order to determine at which stage the release of chlorine takes place, we have transferred pAC25 to *P. putida* strain PRS1 mutants harboring a variety of mutations in the catechol degradative pathway (Figure 3). The results in Table 3 demonstrate clearly that while catechol oxygenase (*catA*), muconolactone isomerase (*catC*) or enol-lactone hydrolase (*catD*) genes are not complemented by the corresponding plasmid-borne genes, the plasmid does produce a gene product that can complement at a reduced efficiency the missing muconate lactonizing enzyme in the mutant PRS2015. Since Reineke and Knackmuss[44] have shown that chlorobenzoate utilizing strains elaborate isoenzymes that are somewhat specific for chlorocatechols, it is not surprising that the plasmid coded 3-chlorocatechol oxygenase does not complement catechol oxygenase. The pAC25-plasmid-coded chloromuconate lactonizing enzyme can, however, act on the nonchlorinated *cis,cis*-muconate, suggesting that chloride release occurs at this stage of the degradative pathway. Whether the chloride release is enzyme-mediated or due to unstable nature of the lactone is not clear at this time.

The lack of complementation of the *catC* and *catD* mutations by the corresponding genes borne on the plasmid pAC25 indicates that either functional *catC* and *catD* genes are not borne on the plasmid, or an entirely different set of enzymes is involved in the biodegradation of chloromuconates derived from 3-chlorocatechol. It is of interest that Evans et al.[45] previously reported the occurrence of a new degradative pathway for the metabolism of 4-chlorophenoxyacetic acid and 4-chlorocatechol by a pseudomonad. The formation of 4-carboxymethylene but-2-en-4-olide as an intermediate of halocatechol metabolism has also been confirmed.[46] According to Evans et al.[45] *cis-cis*-B-chloromuconic acid derived from 4-chlorocatechol undergoes lactonization to

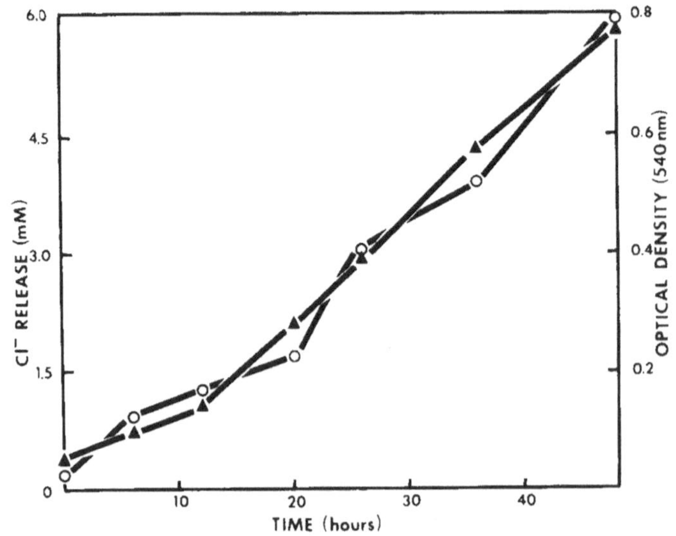

FIGURE 2. Increase in turbidity and chloride release during growth of pAC25⁺ *P. aeruginosa* cells on 3-chlorobenzoate (7.5 mm) as a sole carbon source. ▲—▲, turbidity; O—O, chloride release.

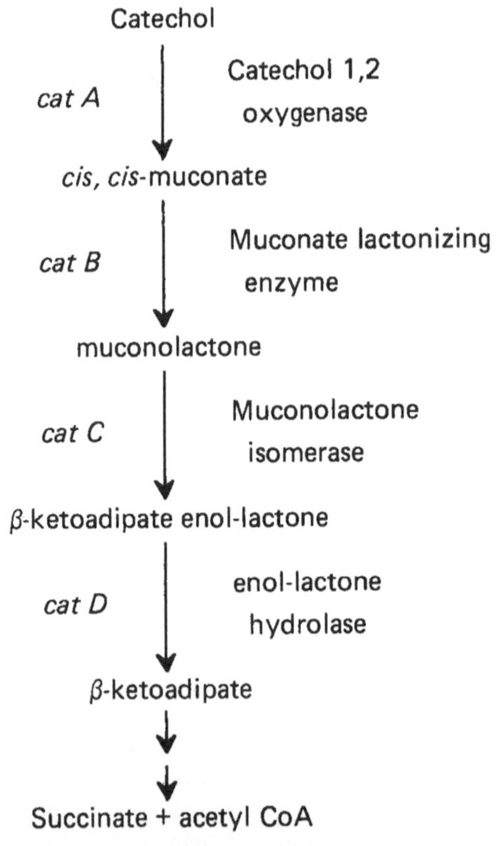

FIGURE 3. Genes and enzymes involved in the ortho cleavage (β-ketoadipate pathway) of catechol.

Table 3
TRANSFER OF pAC25 PLASMID TO CATECHOL-NEGATIVE *P. PUTIDA* STRAIN PRS1 MUTANTS

Donor	Recipient (mutation)	Select	Benzoate-growth phenotype of exconjugants
pAC25⁺Met⁻ *P. putida* PpG1	AC79 (*cat A*)	3 Cba⁺	Benzoate-negative
pAC25⁺Met⁻ *P. putida* PpG1	PRS 2015 (*cat B*)	3 Cba⁺	Benzoate-positive
PAC25⁺Met⁻ *P. putida* PpG1	PRS 2191 (*cat C*)	3 Cba⁺	Benzoate-negative
pAC25⁺Met⁻ *P. putida* PpG1	PRS 5 (*cat D*)	3 Cba⁺	Benzoate-negative

produce γ-carboxymethylene-$\Delta^{\alpha\beta}$-butenolide which is further oxidized via maleylacetic acid. No formation of β-ketoadipate as an intermediate occurs in this pathway. Thus different types of lactonizing and hydrolytic enzymes appear to be involved in the metabolism of chlorophenoxyacetate and chlorocatechols by bacteria.

In order to determine if pAC25 codes for a completely new set of enzymes for the degradation of 3-chlorobenzoate not involving β-ketoadipate as an intermediate, we determined the accumulation of β-ketoadipate in the growth medium of pAC25-positive cells growing with 3-chlorobenzoate or benzoate. When grown with benzoate, pAC25⁺ *P. putida* and *P. aeruginosa* PAO cells produced large quantities of β-ketoadipate as detected by Rothera color reaction, particularly during the exponential phase. In contrast, the same cells produced no detectable β-ketoadipate at any phase of growth when grown with 3-chlorobenzoate. Thus the pathway of metabolism of 3-chlorobenzoate, as specified by the pAC25 plasmid, does not appear to involve β-ketoadipate as an intermediate.

In order to determine if a degradative pathway similar to Evans's pathway with maleylacetic acid as an intermediate[45] may perhaps be specified by the pAC25 plasmid, we synthesized maleylacetic acid and determined the growth rate of pAC25-positive and pAC25-negative *P. aeruginosa* and *P. putida* cells with maleylacetic acid as a sole source of carbon. While the pAC25-negative cells were totally incapable of utilizing maleylacetic acid, pAC25-positive cells grew slowly with this compound.[43] Thus the plasmid appears to code for the degradation of 3-chlorobenzoate by a process similar to the "Evans" pathway.

VIII. CONCLUDING REMARKS: SCOPE OF GENETIC MANIPULATION

The release of massive amounts of toxic chlorinated compounds in the environment appears to have elicited a strong response from the microbial community in the evolution of some novel genetic functions such as the expression genes associated with pAC21 plasmid and a completely new set of degradative genes borne on pAC25 plasmid. The emergence of pure cultures capable of dissimilating such highly chlorinated compounds as PCBs (see Furukawa, Chapter 2) or pentachlorophenol[13-15] clearly signifies the microbial versatility in coping with an extreme environment. Plasmids play an important role in the evolution of biodegradative processes against novel chemicals, as exemplified by the contribution of the TOL plasmid in extending the ability of *Pseudomonas* B3 to utilize 4-chloro and 3,5-dichlorobenzoate. While the plasmid nature of 3 Cba-degradative genes in *Pseudomonas* B3 is becoming clear,[47] a close look at the data of Reineke and Knackmuss[24] shows the absence of the entire TOL plasmid in the 4 Cba-positive and 3,5-dichlorobenzoate positive exconjugants. It is tempting to speculate that the benzoate oxidase gene complex has recombined with a putative 3-Cba degradative plasmid in strain B3 to form a new plasmid that encodes biodegradation of the novel chlorobenzoates. This mechanism of piecemeal assortment of genes in the evolution of degradative plasmids has previously been emphasized.[26]

The involvement of transmissible plasmids encoding dechlorination steps from chloromuconic acid and other aliphatic compounds, may also allow us to construct microbial strains capable of releasing chloride ions from the aromatic nucleus. The emergence of bacterial cultures capable of dechlorinating pentachlorophenol (PCP)[13,14] or 1,2,3-trichlorobenzene[16] has been mentioned before. It would be interesting to find out if pentachlorophenol degradative genes could be carried on a plasmid. It might also be possible to construct strains capable of dissimilating such toxic compounds as 2,3,7,8-tetrachlorodibenzo-*p*-dioxin (TCDD). This can be done by inoculating soil samples isolated from TCDD/pentachlorophenol contaminated areas with pAC25-containing microorganisms in a chemostat. Initially the chemostat may be operated with 3-Cba and pentachlorophenol as sole sources of carbon to allow rapid spread of pAC25 and putative pentachlorophenol degradatives plasmids. With prolonged incubation, the concentration of PCP and 3-Cba will gradually be decreased while the concentration of TCDD will be gradually increased. The rationale is that under increasing toxicity from TCDD and decreasing level of growth substrates, the mixed cultures in the chemostat harboring pAC25 and PCP-degradative plasmids, will be under strong selective pressure to undergo mutations in the plasmid-associated dechlorination genes so as to alter the substrate specificity of the enzymes to include TCDD and its metabolites. This may lead to total or partial degradation of TCDD. It should be emphasized that while the technology for purification of water and to some extent air, is available, there is no technolgy available for purification of soil. The controversy over how to get rid of the toxic chemicals from the Love Canal area or the PCBs from Hudson River sediments, and who should assume the costs, has left the scientific community, the industries, and government regulatory agencies in bewilderment. While the scientific community have successfully monitored the fate and toxicological characteristics of many of these compounds, there have apparently been no attempts to deliberately enhance the capability of microorganisms to release the chlorine atoms from such highly chlorinated molecules to reduce their toxicity. An understanding of the genetic mechanisms of dissimilation of chlorinated compounds, including limitations on biochemical processes and genetic expression, will be very valuable in constructing appropriate strains for enhanced biodegradation and ultimate removal of various synthetic pollutants from the environment.

REFERENCES

1. **Maugh, T. H., II,** *Science,* 202, 602, 1978.
2. **Ames, B. N.,** *Science,* 204, 587, 1979.
3. *Chem. Eng. News,* 57, 14, 1979.
4. **Maugh, T. H., II,** *Science,* 201, 1200, 1978.
5. **Chakrabarty, A. M.,** *ASM News,* 44, 687, 1978.
6. **Dagley, S.,** in *The Bacteria,* Vol. 6, Ornston, L. N. and Sokatch, J. R., Eds., Academic Press, New York, 1978, 305.
7. **Alexander, M.,** in Soil Biology, Reviews of Research, National Resources Research, *UNESCO,* 9, 209, 1969.

8. Siuda, J. F. and DeBernardis, J. F., *Lloydia*, 36, 107, 1973.
9. Kaufman, D. D. and Kearney, P. C., in *Herbicides: Physiology, Biochemistry, Ecology*, Vol. 2, Audus, L. J., Ed., Academic Press, London, 1976, 29.
10. Bourquin, A. W. and Gibson, D. T., in *Water Chlorination: Environmental Impact and Health Effects*, Vol. 2, Jolley, R. L. and Hamilton, D. H., Jr., Eds., Ann Arbor Science, Ann Arbor, Mich., 1978, 253.
11. Hartmann, J., Reineke, W., and Knackmuss, H.-J., *Appl. Environ. Microbiol.*, 37, 421, 1979.
12. Crawford, R. L., Olson, P. E., and Frick, T. D., *Appl. Environ. Micorbiol.*, 38, 379, 1979.
13. Watanabe, I., *Soil. Sci. Plant Nutr.*, 21, 405, 1975.
14. Suzuki, T., *J. Environ. Sci. Health*, B12, 113, 1977.
15. Reiner, E. A., Chu, J., and Kirsch, E. J., in *Pentachlorophenol, Chemistry, Pharmacology and Environmental Toxicology*, Rao, K. R., Ed., Plenum Press, New York, 1978, 67.
16. Marinucci, A. C. and Bartha, R., *Appl. Environ. Microbiol.*, 38, 811, 1979.
17. Goldman, P., in *The Degradation of Synthetic Organic Molecules in the Biosphere*, National Academy of Sciences, Washington, D.C., 1972, 147.
18. Slater, J. H., Lovatt, D., Weightman, A. J. E. Sr., and Bull, A. T., *J. Gen. Microbiol.*, 114, 125, 1979.
19. Perry, J. J., *Microbiol. Rev.*, 43, 59, 1979.
20. Francis, A. J., Spanggord, R. J., Ouchi, G. K., and Bohonos, N., *Appl. Environ. Microbiol.*, 35, 364, 1978.
21. Reineke, W. and Knackmuss, H. J., *Biochim. Biophys. Acta*, 542, 412, 1978.
22. Reineke, W. and Knackmuss, H. J., *Biochim. Biophys. Acta*, 542, 424, 1978.
23. Dorn, E. and Knackmuss, H. J., *Biochem. J.*, 174, 73, 1978.
24. Reineke, W. and Knackmuss, H. J., *Nature (London)*, 277, 385, 1979.
25. Chakrabarty, A. M., in *Plasmids and Transposons: Environmental Effects and Maintenance Mechanisms*, Stuttard, C. and Rozee, K. R., Eds., Academic Press, New York, 1980, 21.
26. Farrell, R. and Chakrabarty, A. M., in *Plasmids of Medical, Environmental and Commercial Importance*, Timmis, K. N. and Puhler, A., Eds., Elsevier/North-Holland Biomedical Press, Amsterdam, 1979, 97.
27. Broda, P., Bayley, S. A., Downing, R. G., Duggleby, C. J., and Morris, D. W., in *Plasmids of Medical, Environmental and Commercial Importance*, Timmis, K. N. and Puhler, A., Eds., Elsevier/North-Holland Biomedical Press, Amsterdam, 1979, 271.
28. Chakrabarty, A. M. and Friello, D. A., *Proc. Natl. Acad. Sci. U.S.A.*, 71, 3410, 1974.
29. Chakrabarty, A. M., Friello, D. A., and Bopp, L. H., *Proc. Natl. Acad. Sci. U.S.A.*, 75, 3109, 1978.
30. Rownd, R. H., Perlman, D., and Aoto, N., in *Microbiology — 1974*, Schlessinger, D., Ed., American Society for Microbiology, Washington, D.C., 1975, 76.
31. Cohen, S. N., *Nature (London)*, 263, 731, 1976.
32. Nakazawa, T., Hayashi, E., Yokota, T., Ebina, Y., and Nakazawa, A., *J. Bacteriol.*, 134, 270, 1978.
33. Jacoby, G. A., Rogers, J. E., Jacob, A. E., and Hedges, R. W., *Nature (London)*, 274, 179, 1978.
34. Bayley, S. A., Duggleby, C. J., Worsey, M. J., Williams, P. A., Hardy, K. G., and Borda, P., *Mol. Gen. Genet.*, 154, 203, 1977.
35. Chakrabarty, A. M. and Gunsalus, I. C., *Mol. Gen. Genet.*, 176, 151, 1979.
36. Deonier, R. C. and Mirels, L., *Proc. Natl. Acad. Sci., U.S.A.*, 74, 3965, 1977.
37. Ribbons, D. W., Wigmore, G. J., and Chakrabarty, A. M., *Soc. Gen. Microbiol. Q.*, 6, 24, 1979.
38. Duncan, C. H., Wilson, G. A., and Young, F. E., *Gene*, 1, 153, 1977.
39. Holsters, H., Silva, B., Vliet, F. V., Hernalsteevs, J. P., Genetello, C., Montagn, M. V., and Schjell, J., *Mol. Gen. Genet.*, 163, 335, 1978.
40. Environmental Protection Agency, *Fed. Regist.*, 31 (143), 30468, 1976.
41. Kamp, P. F. and Chakrabarty, A. M., in *Plasmids of Medical, Environmental and Commercial Importance*, Timmis, K. N. and Puhler, A., Eds., Elsevier/North-Holland Biomedical Press, Amsterdam, 1979, 275.
42. Fisher, P. R., Appleton, J., and Pemberton, J. M., *J. Bacteriol.*, 135, 798, 1978.
43. Chatterjee, D. K., Kellogg, S. T., Hamada, S., and Chakrabarty, A. M., *J. Bacteriol.*, 146, 639, 1981.
44. Reineke, W. and Knackmuss, H.-J., *J. Bacteriol.*, 142, 467, 1980.
45. Evans, W. C., Smith, B. S. W., Moss, P., and Fernley, H. N., *Biochem. J.*, 122, 509, 1971.
46. Sharpee, K. W., Duxbury, J. M., and Alexander, M., *Appl. Microbiol.*, 26, 445, 1973.
47. Timmis, K., personal communication

INDEX

A

B

C

D

E

F

G

H

I

K

Q

R

S

T

W

X

Y